Praise for Stanley Jacobs, M.D.

"It's just like an Indiana Jones movie, but it's all true. The discovery of an Egyptian skin care recipe created thousands of years ago is a real adventure story."

—Benedetta Pignatelli
Vogue Italia

"This is a fascinating story, and it shows that we still have much to learn from our ancient forebears."

—James P. Allen
Wilbour Professor of Egyptology
Brown University

"It is said that all stories are either about the "stranger who comes to town" or "a mystery" or both. While part travelogue, memoir, and a medical treatise on skin care, Dr. Stanley Jacobs' book, Nefertiti's Secret, is primarily "a mystery" story. Firstly, Jacobs searched for copies of several ancient Egyptian Papyri, which he hoped would reveal the secret ingredients used by ancient Egyptians, like Queen Nefertiti, to maintain lovely facial skin tones even while aging. Papyri in hand, he then had to decode the meaning and use of the various ingredients mentioned, especially one derived from hemayet fruit. Secondly, he was the quintessential "stranger who comes to town," as he visits Egypt and other parts of the world."

—Waights Taylor
Author of *Our Southern Home, Kiss of Salvation* and *Touch of Redemption*

"Dr. Jacob's voice is full of humor and has a sense of mystery to be solved. We know that ultimately he does solve the mystery because of the face serum on the market, but this is also a story with history, and not just about chemists in a lab cooking up chemicals hardly anyone can pronounce such as polyethyl. Ohmigod!"

—**Linda McCabe**
Author of *Quest of the Warrior Maid,*
Historian of science and clinical laboratory
Scientist at Healdsburg District Hospital

Nefertiti's Secret

A 3,600-year-old Egyptian papyrus reveals a modern wrinkle cure

By Stanley Jacobs, M.D., F.R.C.S. (C)
with Karen Hart

Map of Egypt

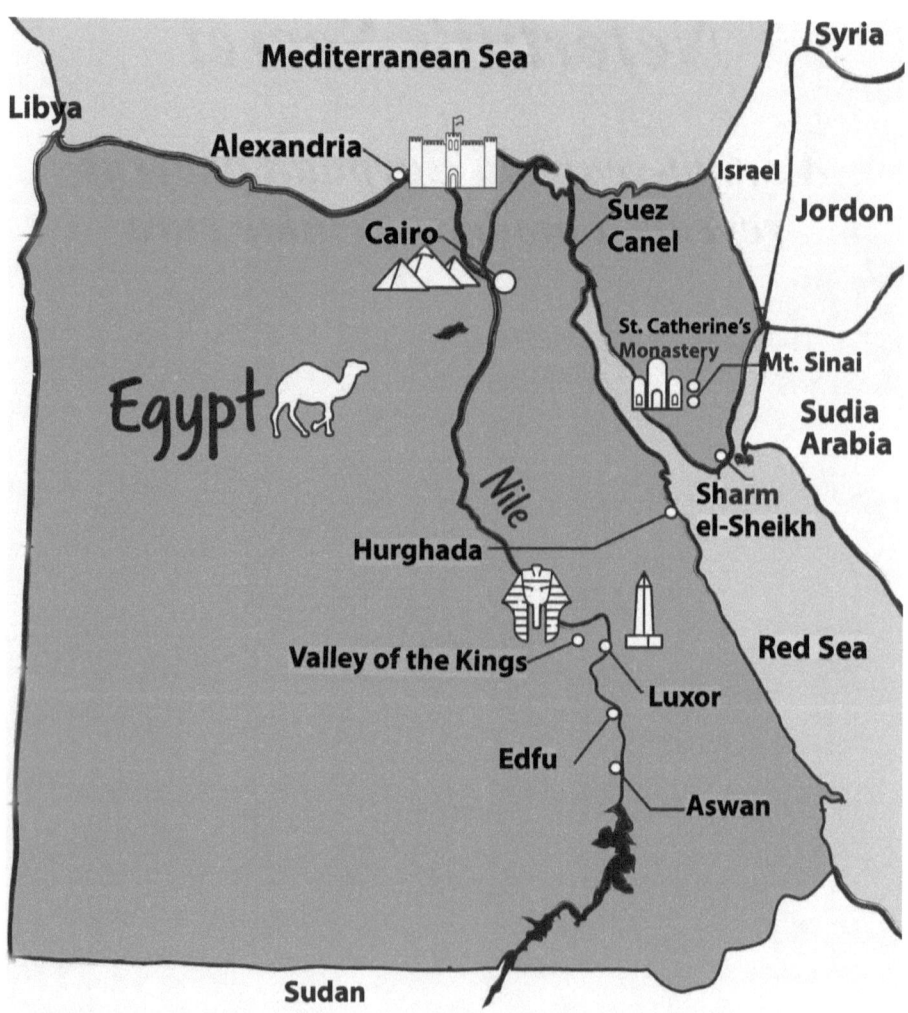

For Joni, Austin and Spencer,
who always put up with my crazy ideas.

ACKNOWLEDGMENTS

First, I would like to thank my wife, Joni, who never ceases to amaze me and for her unflagging support in travelling with me and our sons, Austin and Spencer, around the globe during this quest. I'd also like to thank her for her purist and honest approach when it came to assessing our products. Without that, we would not have produced such genuinely user-friendly products.

I would also like to thank Jules Zecchino for his brilliant cosmetic chemistry mind in developing our products, and Tom Hrubec for his breadth of skin care knowledge and uncanny instinct for bringing people together in a synergistic way. They are both gentlemen who have changed my life as well as my practice of surgery.

Thanks to Karen Hart, a good friend, who has been a gem to work with and a wonderful writer who wove my story together in a manner that flowed beautifully and always maintained the authenticity of my voice. I would also like to thank my staff: Evelyn Mitchell for her tremendous business acumen; Lisa Shields and Lisa "Hillary" Stuckey for their continued sales of our products, and explaining them to our customers; and Lisa Bryant, R.N., who did a great job assisting with my skin care study using the Cutometer.

Grateful appreciation to Dr. James P. Allen for translating the Edwin Smith papyrus, and for his support in editing this book with regards to its Egyptological accuracy.

And finally, I would like to thank my parents, Stanley and Elizabeth Jacobs, for lighting the fire of inquisitiveness and wonder in me. Though they are no longer of this world, I know they are somewhere smiling.

CONTENTS

Introduction

Your Skin Can Rebound and Regenerate

W hat if you could improve the plumpness and elasticity of your skin by 30 percent in one month? This is the first question I ask patients when they want to improve the appearance of their skin. The response, of course, is always a resounding, *"Yes!"* Everyone understands that plumpness and elasticity are key to beautiful skin. In scientific terms, we call this viscosity, the plumpness and elasticity, which provides the snap to skin, or *visco-elasticity*. This is the crux of what matters in skin. It's the difference between a 20-year-old's skin and a 60-year-old's skin.

The fountain of youth is a spring believed to restore the youth of anyone who drinks or bathes in its waters, according to legend. Tales of such a fountain have been recounted across the world for thousands of years. While it certainly would be easier to bathe in a fountain to restore skin, it's not quite that simple and yet it's not impossible to achieve beautiful skin at any age. Skin *can* rebound and regenerate.

Most of my patients are women and their top concerns are to reduce lines, plump wrinkles, achieve smoother skin and reduce brown spots. Global sales of skin care products add up to $218 billion per year. But for years, the feedback I received from female patients was mostly disappointment with the face creams they purchased.

Typically, they've tried scores of products and spent hundreds of dollars on creams that don't work. Still, they persisted with the hope and promise that the latest new cream would be different.

While any lotion can add moisture to your skin and improve the appearance, they often don't work in a significant way. How can that be? Because most skin care companies don't focus on the underlying structure of skin. And they certainly don't look at it from a plastic surgeon's perspective—working with skin surgically, using lasers and chemical peels to enhance the skin's outer layer and working with tissue expanders when necessary. Most cosmetic companies are more interested in the look, feel, scent and packaging of their products. What's more, you may be surprised to know that the United State's Food & Drug Administration (FDA) doesn't get involved with the cosmetics industry the way they do with the pharmaceutical industry. The FDA only requires that cosmetic companies not use chemicals from a banned list of ingredients. As a result, well-known brands don't have to prove anything other than marketing works to turn profits, while lesser-known brands must prove themselves, or no one would purchase them. But in either case, there's usually no real science behind these products, which is why they don't work. They're simply selling a moisturizer to treat the symptoms of dry looking, wrinkled skin. (You could use olive oil for the same outcome and it would work, though your skin may feel greasy and you end up smelling a little like salad dressing.) It's a short-term fix, much like wearing a little more makeup to give the appearance of looking rested when you didn't get a good night's sleep. And once you wash the makeup off, your skin looks the same as it did before using it.

A good skin cream should treat the underlying structure of your skin, and that's why I was reluctant for years to recommend or manufacture a product for my patients. As a board-certified facial plastic surgeon, I was only interested in a product that could scientifically live up to its claims. I've been helping patients improve the appearance of their skin for more than 30 years through surgery and chemical peels, which lead to dramatic changes. But for years I've been searching for a product that my patients could use at home to maintain their skin before and after treatments.

The science presented itself to me one day, unexpectedly and by accident, when I happened upon a book, a medical treatise, used by Egyptian physicians 3,600 years ago. It was a rare find and having a passion for their culture and an appreciation for the surgical methods used in ancient times, I was intrigued. Initially, the formula seemed to contain a primitive ingredient from ancient times, but after working with it in a lab with a cosmetic chemist we discovered that it had a surprisingly sophisticated composition that is scientifically proven to improve the plumpness and elasticity of skin.

The Egyptian culture is steeped in preservation and making the skin look good. When I first read the recipe, I thought it would be simple, but in reality, it was an elaborate formula. When you consider that Egyptians lived in the desert and were always in the sun, it makes sense that they wanted an unguent (or ointment) that worked. The value of this recipe is obvious because the original recipe instructed that it be placed in a vase of costly stone such as lapis lazuli, jasper or alabaster. In the days of tomb raiders, the first items stolen were the lotions, creams and unguents, not the gold. Why did they bother with these items? The Egyptians were master chemists and physicians; they were geniuses and highly advanced. And so perhaps that's why the lotions were claimed first.

Is there a fountain of youth? Perhaps the legendary fountain isn't a fountain at all, but an unguent, originated by Egyptians. Queen Nefertiti is an icon of beauty used as a standard by most facial plastic surgeons because of the dimensions of her face. Her name "Nefertiti" means beauty has come. This book tells the story of how I discovered a recipe for a face paste written on a 3,600-year-old papyrus by a scribe and overlooked by medical science for thousands of years. It describes my journey to create a serum from the recipe that lives up to its claims, but it is also my hope this book will serve as your guide to helping you achieve beautiful skin. Did Nefertiti once use this serum? I think so. Perhaps this is her parting legacy to the modern-day world.

Stanley Jacobs, B.Sc, M.Sc,M.D.,F.R.C.S. (C)
Facial Plastic Surgeon
Director of The Jacobs Center for Cosmetic Surgery
CEO - World Skin, LLC
San Francisco and Healdsburg, California

Part I: Ancient Wisdom ~ Modern Thinking

"...no professional owes more to the long-ago past than does the doctor."

—*Otto L. Bettmann*

Chapter I

The Egyptian Connection

T here was never one defining moment in my quest to develop a face serum. In retrospect, however, there was a consistent ebb and flow of clues and events that propelled me along, beginning with childhood.

As a young boy growing up in Canada, I admit hockey was my foremost passion, but I was also captivated by ancient Egypt, the pyramids, the mummies, the discovery of King Tutankhamen's tomb, the legend of Cleopatra. My aptitude for science took hold later as a teenager when I attended the English-speaking Riverdale High School in Pierrefonds, Quebec. There, I came to understand the profound role ancient Egypt played in developing some of the earliest-known forms of the sciences, which included the field of medicine, my chosen career path.

Though my interest in ancient Egypt never completely left me, it was put on hold as I completed an undergraduate degree in science, and a graduate degree in medical biophysics, which later played an important role in my scientific endeavors. I went on to medical school, completed an internship, residency and fellowship, and then established my own practice as a facial plastic surgeon in Northern California.

In March 2000, I was firmly planted in mid-life, married with two young sons, when a chance encounter with a new patient triggered my interest in ancient Egypt once again. At the time, it seemed nothing more than a pleasant exchange, but the patient was key in fueling my interest in ancient Egypt and would eventually become a friend. A pathologist from New York City had just moved to Santa Rosa and had made an appointment at my office. We were in an exam room, and I was going over his file, jotting a few notes and getting acquainted and discovered that we both shared a passion for ancient Egypt. He told me he was a member of the American Research Center in Egypt, known as ARCE. (ARCE is a professional organization that supports archaeology, scholarship training and conservation efforts in Egypt.)

"The annual meeting is coming up. Next month," he said. "And this year it's at U.C. – Berkeley."

I considered him for a moment. Berkeley was only an hour or so away. It seemed more than a mere coincidence, and I instinctively knew I should go.

Given that medical matters were of keen interest to both of us and we shared an interest in ancient Egypt, I asked him a question that had been on my mind for years. "Do you happen to know of any books that detail ancient Egyptian surgery?"

"*The Edwin Smith Surgical Papyrus*, written by James Henry Breasted," he said, without a moment's hesitation.

It had never occurred to me that I needed to research papyri. I jotted the information down on a Post-it note and slipped it in my pant pocket. It wasn't likely to be light reading and definitely not the sort of book to make the *New York Time's* best-sellers list, but I knew Egyptians had practiced some of the earliest-known human surgeries, and I wanted to explore the topic in more depth. This seemingly random exchange with a new patient would become the first step of a remarkable journey that would consume me for the next eight years.

Chapter II

A Rare Find
April 2000~ARCE Meeting
University of California, Berkeley

I spent most of the day at the ARCE meeting, listening to various speakers discuss ancient Egyptian life. It was like nothing I had attended before, and I was pleasantly surprised to find the same academic rigor biological scientists apply to their research was applied in Egyptology. Afterwards, I wandered over to the book vendors. I had the Post-it note in my pocket. Naturally, I'd looked the book up on the Internet, which was not what it is today. The information was sparse at that time, so I had no idea what the book looked like. All I knew for sure was that it wasn't available online.

There were rows of tables piled with books and about a dozen vendors lining a hallway. The distinctive scent of old books left in an attic over time, permeated the air. I've always enjoyed reading, and old books have a scent that signals a text's authenticity to me on a visceral level. I moved from table to table, asking the same question: "Do you have a copy of the *Edwin Smith Surgical Papyrus?*"

The response was always the same. No. They either didn't carry the set, or they didn't have it at the time, but I learned more about the book from each vendor I spoke with.

"Only 200 to 300 copies of the first edition are thought to exist," said one vendor.

"Very hard to find," said another vendor.

"Locating the set at an ARCE meeting would be unusual," advised another vendor.

I was growing more discouraged with each vendor I spoke to, realizing that finding the book would be a long shot. Still I persisted, stopping to ask each vendor and learned more along the way. The "book" as it turns out, was a rare, two-volume set. One volume contained the original hieroglyphic writings; the other included a translated text. The book had been published in 1930, and no one had bothered to reprint it in the last 70 years. I understood this isn't the sort of book that everyone wants to read, but it would seem that there had to be other surgeons out there who would find the set of interest. Years later, I found that there are many surgeons interested in historical books about surgery, yet no one I'd ever encountered had ever heard of the *Edwin Smith Surgical Papyrus*.

By the time I'd stopped to talk with about 10 vendors, I knew it was highly unlikely I'd find it that day. There was one vendor left, and I wasn't getting my hopes up, but I approached his table and glanced at the books, recognizing some of the same titles and covers from previous vendors. It was a long shot, but I inquired anyway.

The vendor was a slight man with a diminutive build. "Yes," he said, in a distinct New Jersey accent.

Yes? I wasn't expecting that. Clearly, there was a misunderstanding. Did he mean that yes, he had heard of the book, but of course, he didn't have it? I asked again, "You have the Edwin Smith Surgical Papyrus? The two-volume set?"

"Yes. You're in luck. I have only one set, and I don't usually bring it to events. They're scarce and fairly precious and can't withstand the travel. But this time, I just happened to pack them anyway." He shrugged, and then turned and bent over the boxes on the floor behind him.

I felt my heart hammering in my chest, as if he was about to produce *The Dead Sea Scrolls*. Though truth be told, even if he had scrolls that day, I would've preferred to get Breasted's two volumes. The vendor pulled the two large, rare books from a box on the floor.

I couldn't believe my good fortune. The vendor set the books on the table with care. They were larger than I'd expected. The covers were an unremarkable tan woven fabric. I picked up each book separately and thumbed through the pages. They were slightly frayed at the bindings, and the edges of the pages were a bit worn, but, overall, they looked to be in excellent condition. The first book, which was larger, was only about a half-inch thick but contained photographs of the papyrus; it was sun-bleached, leaving a phantom shadow where its companion had stood on the shelf for years. The second book, was somewhat smaller, but weighed a great deal more and was almost three inches thick; it contained the translation of the hieroglyphs into English.

I looked at the vendor. "How much for the set?"

"Four-hundred and fifty dollars."

Ouch. It seemed a steep price, but they were rare and I knew I was unlikely to find another set anytime soon, if ever. I reached for my wallet and we made the transaction.

As I left the building that afternoon and made my way across the campus parking lot to find my car, I felt like Indiana Jones in the opening scenes from *Raiders of the Lost Ark*, escaping the cave after tipping the golden idol off its stand. I was half expecting a giant stone orb to roll after me on my way through the parking lot. (Of course, the irony is now you can go online and instantly order a paperback version of the same set and qualify for super saver shipping, but at the time it was a rare find.)

I unlocked my car and set the two volumes on the passenger seat. I couldn't wait to show my wife, Joni, and two sons the books, and share the story of how I happened to find them. I felt inspired. I was looking forward to learning about the surgical methods used in ancient Egypt.

While I drove north on Highway 101 to my home in Sonoma County, I glanced at the books beside me from time to time. I caught myself speeding more than once and had to slow down. While I was still marveling over my astonishing good luck in finding this rare set of books, I had no idea they contained a sophisticated formula that could revolutionize skin care as we know it today. And it had gone unnoticed by medical science for thousands of years.

Chapter III

The Lost Elixir

T he two-volume set of the *Edwin Smith Surgical Papyrus* was kept on the coffee table in our living room. And though Austin and Spencer were still young and impulsively playful and mischievous at times, they both knew not to get carried away with their horseplay around the books. In the months that followed, I began reading the English translation in the evening, slowly and with no specific intent other than for pleasure. This volume was nearly 600 pages, and I intended to take my time and enjoy it.

As I began reading, I found that it was mostly a surgical treatise about traumatic wounds from battles and treatments for the wounds. What's particularly fascinating about it is that it is the *first* medical treatise, describing a diagnosis followed by the method of treatment in recorded human history. This predates Greek physicians such as Galen and Hippocrates by about 1,000 years. I was immediately struck by its scientific integrity. The treatise was factual, surprisingly sophisticated and there were no pretenses.

Egyptian doctors knew what they could and *could not* accomplish during those times. When there was no known medical treatment, opium or alcohol was administered to help ease any pain. When no treatment was known, the scribe wrote: "This is an

Did you know?

Over a period spanning three millennia, the ancient Egyptians became master chemists and doctors who developed drugs and performed some of the earliest-known surgeries. Their expertise was recorded by Homer in *The Odyssey*, and the skill of their midwives referenced in the Bible (Exodus 1:19). Hippocrates—known as the Father of Western Medicine—also praised the knowledge and expertise of ancient Egyptian physicians in his extensive writings during the third century B.C.

ailment we have no treatment for." I found this so refreshing since so often physicians want to occasionally treat medical conditions, even when we're unsure of the cause or if the treatment will work.

"Ancient Egyptians were clever and advanced in their techniques," I said to Joni one night, as we were reading together in the living room.

Sometimes I read passages to her. Given her own medical background as a nurse, she found the book interesting as well. Many of the procedures described were treated in much the same fashion as physicians practice in the modern-day world. A nasal fracture, for instance, was set in much the same way it is today. The Egyptians used seeds of grain for nasal packing and bandaged the nose externally nearly the same as we do now. Though doctors use more advanced materials such as intra-nasal splints or casts made of plastic, the procedure performed for a nasal fracture was basically the same. Ancient Egyptian doctors also performed a procedure called "brain trephination," which involves drilling into the skull to relieve a blood clot around the brain, known as a *subdurmal hematoma* in medical terms. This is caused by head trauma and if not drained it will result in death. We still perform the same maneuver today, the only difference is our electric drills have a self-stopping mechanism when it reaches the *dura*.

As a facial plastic surgeon, I was fascinated with the book, because most of it contained procedures for the face—the nose, the lips and the eyes. There were hundreds of treatments, and it appeared to be the work of a single scribe, as dictated by a single surgeon.

One summer evening, I settled in

my chair with the book, turned the page and found a section titled: "Recipe for Transforming an Old Man into a Youth." Breasted notes the recipe is written in a different handwriting at a later date. It was as if the initial scribe put his pen down after describing surgical procedures for the thorax. The text simply stops. It's one of the mysteries of the book.

The title immediately impressed the hell out of me. I could understand their obvious desire to combat sun damage, given that they lived in the desert, but this "recipe" was much more than that. The point of it was *to remove wrinkles* and *all signs of aging*. And because the entire text had medical integrity, I was immediately intrigued. Ancient Egyptians apparently valued skin perfection as much as we do today. I began reading:

"Let there be brought a large quantity of hemayet fruit, about two khar [bushel]… It should be bruised and placed in the sun. Then when it is entirely dry let it be [husked] as grain is [husked], and it should be

About James Henry Breasted

Born in 1865, James Henry Breasted attended Yale University, following some prior studies in pharmacology and a brief stint working as a pharmacist. Transferring to the University of Berlin, he received his doctorate in Oriental Studies at age 29, specializing in ancient Egypt. A year later, Breasted was appointed to the faculty of the University of Chicago, the only American institution that offered courses in the burgeoning field of Egyptology at the time.

In addition to being knowledgeable about the languages of ancient Egypt, Breasted was thorough in his work. He often called upon other renowned Egyptologists for help with translations, and frequently referenced works containing similar hieroglyphic fragments. Breasted also did not make leaps of faith. If he was unsure about the meaning of a word, he said so, and was clear about what was speculation on his part. Breasted was so revered in his field of hieroglyphic translation that when Howard Carter made his legendary discovery of King Tutankhamen's tomb in 1922, it was Breasted he summoned to perform the necessary deciphering work.

winnowed until (only) the fruit thereof remains."

I continued reading with great interest, but what captivated me was the final paragraph:

> *"Anoint a man therewith. It is a remover of [wrinkles] from the head. When the flesh is smeared therewith, it becomes a beautifier of the skin, a remover of [blemishes], of all [disfigurements], of all signs of age, of all [weaknesses] that are in the flesh. Proven good a million times."*

The excitement building inside of me ignited a flame and continued to grow stronger. *Could this be real? An archaic elixir to remove wrinkles?* I re-read the seven pages of translation over and over that evening.

There were two other skin-improvement recipes on the left-hand side of the book in the verso. But these recipes were brief, consisting of simple combinations of honey, red natrol (a carbonate salt used extensively in mummification), salt and calcite powder. They But once again, the scientific authenticity struck me. Both recipes promised nothing more than to "beautify the complexion." The salt grains served as an exfoliant, and the honey provided antibacterial properties, both justifiable ingredients that would produce the desired results. So reading a recipe to remove wrinkles, well, I was consumed with interest.

Here I was at the crest of a new millennium with an ancient formula that might possibly live up to its claims. The Egyptians were advanced and I was curious. That evening I knew I wanted to re-create the formula. My family had long since turned in for the night when I finally closed the book and set it on the coffee table. *A remover of blemishes, disfigurements, of all signs of age? Could this formula really work?*

The next morning at breakfast, I was showing the recipe to Joni and the boys when I was reminded that I'd need to decode a key element—the sole ingredient, in fact. Hemayet.

What was hemayet?

I'd been so caught up with the possibilities of the formula the night before that I hadn't yet searched online for an answer, but I wasn't concerned. Breasted was unable to decipher the meaning of this word, but that was 70 years ago. He wrote: "It evidently had a hard husk or shell, which suggests that it might even have been a variety of nut."

Breasted was considered the authority of his time when it came to deciphering hieroglyphs and he was thorough in his work. If he was unsure about the meaning of a word, he said so and was clear about what was speculation on his part. It seemed extremely odd that with all his understanding of hieroglyphs, he wouldn't know this one particular and important word. And if Breasted didn't know what hemayet was, then who would? Nevertheless, 70 years had passed since he'd translated *The Edwin Smith Surgical Papyrus*. Surely an Egyptologist had translated the term during this time.

Later that day, I searched online, typing in "hemayet fruit." Nothing. Was it spelled differently? I tried "hemahyet." Again, nothing.

I returned to the recipe and re-read the directions, searching for clues.

"Let it be set aside,

Simply a Face Paste?

Ancient Egyptians apparently valued flawless skin as much as we do in the 21st century. This makes sense when you consider that temple walls always portray youthful-looking people. Their face and skin, as far as I know, are always beautiful—there are no teenagers plagued with acne or Egyptian elders portrayed with wrinkles or brown spots. But what's humorous about James Henry Breasted's translation is his reaction to the recipe. He sounds somewhat annoyed and dismissive of the resulting formula when he writes: "…simply a face paste, which will remove wrinkles."

In the translation, he clearly hoped this recipe was a fountain of youth, Ponce-de-Leon-type of thing. He dismisses it as a superficial topical application, and not something an aging person would drink as a way to magically reverse the biological clock from the inside out. In all fairness, Breasted was working on the project of his lifetime, translating a document that could possibly be the greatest contribution to mankind's knowledge of early surgery. Given his scholarly background, The *Edwin Smith Papyrus* must have seemed like a treasure trove of medical secrets, especially if he'd translated the recipe for an elixir for everlasting life. But a face paste? Whatever. Not so amazing.

mixed with water. Make into a soft mass and let it be placed in a new jar over the fire and cooked very thoroughly, making sure that they boil, evaporating the juice thereof and drying them, until it is like dry, without moisture therein. Let it be dug (out of the jar). Now when it is cool, let it be put into (another) jar in order to wash it in the river. Let it be washed thoroughly, making sure that they are washed by tasting the taste of this water that is in the jar (until) there is no bitterness at all therein. It should be placed in the sun, spread out on launderer's linen. Now when it is dry, it should be ground upon a grinding mill-stone."

Was it the cooking process that made the fruit bitter, or did it taste that way? Grapefruit initially came to mind, but that was certainly wrong. It was too large and didn't have a husk, so it couldn't be processed like grain.

In the weeks that followed, I began to wonder if this fruit was still around. And if it were around, would it still have the right genomes or chemical characteristics needed for the recipe to work today? All living species of plants evolve over time to some degree. This is especially true when humans cultivate plants.

One day I reread the final process described in the recipe, looking for clues.

"Let it be set aside in water. Make like a soft mass and let it be placed in a jar over the fire and cooked thoroughly, making sure that it boils, that the fluids of the mass may go forth therefrom. A man shall dip out the mass that has come of it with a dipper. Put into a hin-jar after it is [of the consistency] of clay. Rub and make thick its [consistency]. Dip out this mass and put upon a linen cover on the top of this hin-jar. Now afterward it should be put into a vase of costly stone."

I found nothing in the way of clues, but the ointment must have been considered valuable, or the ancient Egyptians wouldn't have wasted time or resources putting it in an expensive container. It was yet another intriguing aspect about this ancient recipe that captivated my imagination and drove me to find answers for this lost elixir.

About the *Edwin Smith Surgical Papyrus*

Today, the *Edwin Smith Surgical Papyrus* is owned by the New York Academy of Medicine. One of the primary reasons the *Edwin Smith Surgical Papyrus* included the word "surgical" in the title is because it was unlike other known medical papyri, as it was free of any spells or incantations. In fact, only in one passage, "Forehead Wound with Skull Fracture," is a spell administered along with the prescribed medicine. The passage reads:

> *"The enemy in the wound has been driven off; the conspiracy in the blood has been made to tremble; the vulture of every side has been given to the mouth of the effective goddess. This temple will not deteriorate; there is no crocodile or poison therein. For I am in the effective goddess's protection: Osiris' son is rescued."*

When I read this passage, the language immediately struck me. Yes, there technically is an enemy in the wound—and it happens to be known as what we call "bacteria" today. And depending on the type of infection, that enemy may very well launch a form of bacterial conspiracy against the unlucky host.

Bottom line: behind this poetic incantation was a fundamental truth. And when modern-day physicians encounter infections or other conditions that don't initially respond to a prescription or other treatment, they *try* another cure and *hope* it works — just one reason why it's called "practicing medicine." Given the complexities of the human body, a one-for-one relationship for treatment and cure is not always realized, which is why many surgeons silently hope for a positive outcome. Or, if the surgeon is religious, it's not uncommon to say a prayer before an operation (and encourage the patient's family to do the same). What's more, many hospitals also have chaplains and chapels to accommodate a variety of faiths. Prayer may not be formally documented or regularly practiced, but the custom of calling on a higher power for help in healing has not entirely vanished from today's medicine. Belief can be a powerful force, and there is evidence to suggest that ancient Egyptians understood the role it played in treatment.

Dr. Stanley Jacobs with the Edwin Smith Papyrus.

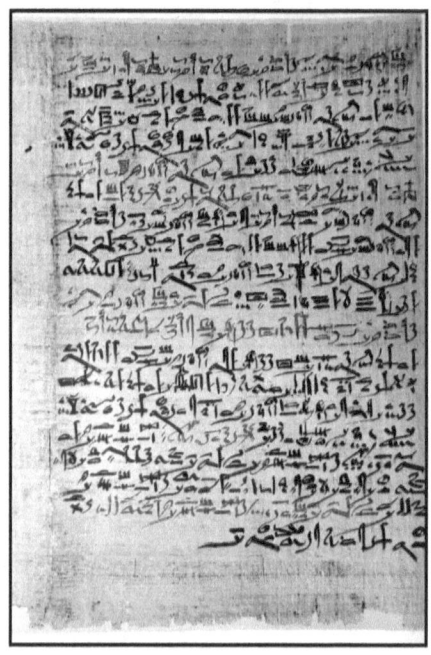

**Edwin Smith Papyrus,
Hieroglyphs**

**Edwin Smith Papyrus,
Hieratic**

Chapter IV

The Odyssey Begins

I called several Egyptologists at the University of California, Berkeley, but they were also mystified by the word "hemayet." My next step was to hit the books. I had a fair number I'd collected over the years on ancient Egypt, but most covered the usual topics in broad terms such as the pyramids, mummies, pharaohs and civilization. There was not much information in the way of medical practices or terms, let alone the ingredients used to make remedies. I scoured the bookstores in the San Francisco Bay Area over the next several months and felt as if I'd visited them all, but my "hemayet" search revealed nothing. In fact, the term "hemayet" was never mentioned in a single book I found.

I did, however, find a few excellent texts that broadened my knowledge of ancient Egyptian medicine, but one text in particular suggested that searching for ancient Egyptian terms could prove to be a waste of time:

"Frequently, especially in medical texts, it is impossible to identify the ancient Egyptian word for a known disease or ingredient (such as an herb or plant) particularly when it is not described. For the ancient Egyptian reader of such texts detailed description of such things was unnecessary as they were familiar items."

The Ebers Papyrus

The *Ebers Papyrus* (also known as the *Papyrus Ebers*) is an Egyptian medical papyrus of herbal knowledge, which was written in about 1500 B.C. It is among the oldest and most important medical papyri of ancient Egypt and was purchased in Luxor in the winter of 1873-74 by Georg Ebers. Currently, it's kept at the library of the University of Leipzig in Germany.

The *Ebers Papyrus* is written in hieratic Egyptian writing and preserves the most voluminous record of ancient Egyptian medicine known. The scroll contains 700 magical formulas and remedies.

It made total sense. There is no need, for example, to define "baking powder" in today's recipes because it's widely understood. But 3,000 years from now, who knows? Baking powder might fall out of favor to a more sophisticated ingredient, or take on a catchy new name with few references left to explain the older terminology.

The more I read, the more I came to understand that ancient Egyptians were as concerned as we are about the visible changes to the body brought about by aging. For example, several recipes in the *Ebers Papyrus* address graying hair, and one formula supposedly grew hair on a bald head. Ingredients in the latter recipe included dog bone and donkey's hoof among other substances of questionable effectiveness. On the flip side of the self-improvement coin, a recipe containing lotus leaves could be applied to the scalp of a "hated woman" to make her hair fall out. (That's pretty creepy.)

Several of the medical papyri include recipes that were attributed to the gods themselves, which would have boosted the credibility of a prescription and given patients a higher level of confidence in its effectiveness. "It would be quite wrong to dismiss magic as irrelevant to the healing process," writes John F. Nunn in *Ancient Egyptian Medicine*. "Suggestion and expectation of a cure have a measurable curative value, particularly in the relief of pain, a phenomenon now known as the placebo effect."

Given what they understood, the ancient Egyptians did a commendable job of sorting out which illnesses and conditions they could address with treatment, and which they sought help for through the spiritual realm.

Still, it was hard not to be amused sometimes by a few of the other recipes I was learning about in the course of my initial research. In the *Ebers Papyrus* were remedies "to expel an evil inflammation," and to treat "a pain in the anus of demonic origin." (When you've got one, it can certainly feel this way.) The remedy to dispel this devilish pain was a drink made from wormwood, juniper berries, honey and sweet beer. The alcohol would've helped alleviate the pain at least.

In the *Chester Beatty Papyrus* there was a recipe to treat yet another patient sidelined by an aching anus: "Knead with honey burnt wolf's dung ground with white pepper. Let the patient drink it. But first claim your fee. A proven remedy."

I wasn't sure whether the doctor's need for payment in advance was because the dung had knocked off a few people here and there, or because collecting his fee was a problem in ancient times. Neither proposition seemed particularly encouraging. But the last statement —a proven remedy—caught my attention because it reminded me of something I read in the formula for transforming an old man into a youth. *"Proven good a million times."*

The phrase kept ringing in my head. And it was one clue that led me to believe the recipe might be more about science than a supernatural invocation.

Chapter V

The Early Days
Montréal~Northern California
Medical Training 1980s–Mid-1990s

M y passion for medicine began as a young man. I attended McGill University in Montréal where I studied anatomical science, majoring in immunology and researching lymphocytes, a type of white blood cell. Once I finished my bachelor's degree, I completed my master's degree in medical biophysics, with an emphasis on T-cell immunology at the Ontario Cancer Institute in Toronto. I was then accepted into the University of Western Ontario Medical School in London, Canada.

I realized fairly quickly that I had an inherent instinct for surgery and thought I might combine my background in immunology with surgery, meaning I was headed toward a medical career in transplant surgery. Later, I worked as a thoracic surgery resident at Mount Sinai Hospital in Toronto and later became proficient at anastomosis, the process of connecting arteries, which is the bread-and-butter of cardiac and vascular surgery, at the Toronto General Hospital. Along the way, I realized I had a natural aptitude for this work, but there was little creativity. As a surgeon, putting an artery where you think it would look better aesthetically is hardly a good idea.

While I was there, I met a young woman who worked as a thoracic specialty nurse. One autumn morning, which happened to be

The World's Most Beautiful Woman

Photo by Joni Jacobs

Her left eye is missing, and both ears are damaged by time. She is more than 3,000 years old, yet her preternaturally elegant face continues to spellbind us. She is Neferneferuaten Nefertiti—more commonly known as Nefertiti, the great royal wife of the Pharaoh Akhenaten. With her husband, she reigned what was arguably the wealthiest period of Egyptian history. Her name means beauty has come.

While Cleopatra is considered an ancient beauty and may be the more well-known Egyptian queen, Nefertiti was by all accounts the more beautiful. An ancient sculptor captured her graceful features for eternity on a 17-inch-tall bust of such astonishing beauty that it became even more fa
mous than her name. To this day, many facial plastic surgeons use a copy of Nefertiti's three-dimensional bust in their practices as a reference, including me.

Nefertiti's almond-shaped outer eye angulation, heightened curved cheekbones, and lip fullness are the key elements to female facial beauty. Why is her face shape considered the gold standard for female facial harmony? Is it because she has epitomized beauty for so long? If so, then why is she still considered beautiful by modern-day standards? And who made up the rules on attractiveness and when?

Scientific studies on female facial proportions now help explain this conundrum. But even if Nefertiti's beauty was exaggerated on this legendary bust, then an ancient sculptor hit an anthropomorphic home run by portraying a virtually perfect female face that portrays timeless beauty.

my first day of surgical rounds at the hospital, I was running late and sprinted through the hospital corridor to the nursing station on the 14th floor. It was just after seven o'clock. "Where's the team?" I shouted, and though I didn't realize it in the moment, my inquiry came out all wrong, rushed, perhaps abrasive and definitely discourteous.

She looked up from the patient chart she was working on, regarded me for a moment, and then said in a clipped even tone, "If you'd been here on time, then you'd know." Each word was embedded win ice, and the look on her face said it all. *Here's yet another rude surgical resident to deal with for the next three months.*

"I'm sorry." I smiled at her, hoping to remedy the situation quickly, but she didn't smile back. She considered me only briefly, before returning to her chart. "Down the hall, last room."

"Thanks," I said, appreciating her cheeky spirit and blue eyes. She was blonde, fair and slim.

In the days that followed, I learned her name was Joni Peppler. Eventually, I worked up the nerve to ask her out to amend for my rudeness that day, but I admit she had captured my attention, and I wanted to get to know her. We made a date for November 7, 1984, to attend a musical venue with her friends at seven o'clock and I made sure I arrived on time. Apparently, I fully redeemed myself because we dated exclusively after that evening.

As our relationship unfolded in the months ahead, I also took advantage of a three-month rotation in ear, nose and throat, which included facial plastic surgery. I quickly found that in one day, I might reconstruct an ear, sculpt a nose, remove a cancer of the thyroid gland, or repair damage to the neck caused by a knife wound—all of which require various techniques and procedures. I enjoyed the challenge and the variety of these types of surgeries. After two years of general surgery, I made a decision to pursue ear, nose, throat and facial plastic surgery at the University of Toronto. As a result, I found myself increasingly appreciating facial aesthetics—an ethereal quality that supposedly defies quantification, though some facial plastic surgeons make it their business to scientifically measure facial dimensions.

As a resident in facial plastic surgery, my work was demanding, but four years later, Joni and I were married on a beautiful, sunny June day

A Defining Moment

My interest in skin took hold when I was an ENT/facial plastic surgery resident in Toronto. (ENT means specializing in Ears, Nose and Throat.) A young male patient I was helping had a large, brown furry growth, covering most of his nose. Naturally, its appearance was his main concern, but there was a small chance it could be malignant and he wanted it removed.

Because the growth was so large, nearly all of his nasal skin had to be replaced. First, we used a tissue expander to stretch the skin on his forehead, so we would have enough skin to cover his nose. Once the skin had sufficiently expanded, we were able to surgically make a midline forehead flap. This skin of the forehead was lifted up and brought down to cover his nose. Because this young man had prime visco-elasticity to his skin, his surgery was a success and eight weeks after surgery, you'd never guess he underwent surgery. If we had performed the same procedure on a 65-year-old, it wouldn't have been as successful, due to poor visco-elasticity.

in 1988 in Barrie, Ontario. Though my interest in Egyptian culture and history was still somewhat dormant in those days, my fascination never left me and wouldn't fully resurface for 12 more years.

After I finished my facial plastic surgery residency in Toronto, I was accepted into the University of California, San Francisco (UCSF) fellowship program, which is an extra year of training after residency. I was honored to work alongside Larry Schoenrock, M.D., who was instrumental in building the society known as the Foundation for Facial Plastic Surgery. During my fellowship, I performed surgeries at both UCSF and in Santa Rosa, which is about an hour's drive north of the city.

Though we're both originally from Canada, Joni and I fell in love with the San Francisco Bay Area and decided to stay. In 1991, I joined a practice—specializing in ear, nose and throat and facial plastic surgery—in Sonoma County. I opened a facial plastic surgery office in Santa Rosa in the mid-'90s, and later opened an additional office in San Francisco as my practice grew.

In the years that followed, we had two sons, Austin and Spencer. At mid-life, Joni and I were raising our sons, who were in elementary school at the time. My practice was expanding and I was thinking about adding a line of skin care products for my patients, but it was mostly a vague notion. I was first and foremost a facial plastic surgeon, working exclusively on the neck up, and I knew very little of cosmetic chemistry. What's more, I knew there was already a multitude of skin care products on the market and that many of them didn't produce the results women wanted. I had no interest in producing another me-too product that failed to deliver on its claims.

Chapter VI

In Search of Hemayet

A s I continued to read Egyptian texts, I soon discovered that the "Recipe for Transforming an Old Man into a Youth" was not the only formula claiming repeated success. At the end of several recipes in the *Ebers Papyrus* asserting repeated success, were comments such as: "Really excellent," or "Really efficient, tried a million times."

And in a later text, I found this: "God knows that this recipe is a good one!" This claim was made about a dill mouth-rinse that must have been about as spectacular as a mouth-rinse could have been for the time.

Today, before a new drug can be sold to the public, it must undergo a series of scientific tests and clinical trials that typically take years to complete. The rules for today's shampoos, cosmetics, skin lotions and other non-pharmaceutical products are less rigorous, but only because their individual ingredients have passed muster in previous testing, or otherwise proven safe through long-term, widespread use. Ingredients such as glycolic acid, isotretinoin and hyaluronic acid are proven to be safe. Paraben, however, is an ingredient you want to avoid.

The value in scientifically testing a pharmaceutical product is to eliminate risk, which in turn engenders patient trust, and it appears the ancient Egyptians understood this concept. As these recipes passed through the hands of ancient Egyptian physicians over the centuries, there was plenty of time available to observe their efficacy. Almost 300 years separate the *Ebers Papyrus* and the *Chester Beatty Papyrus,* for example, so the effect of any formulas that survived from that one document to the next had been observed in patients longer than the United States and Canada have been in existence. Even more incredibly, the total time between the earliest- and latest-known Egyptian medical papyri spans an astounding two millennia—the total time since the birth of Jesus. This incredibly long duration of Egyptian civilization and know-how, is what I still can't get my head wrapped around. They go back farther B.C., than we go forward A.D., by 1,000 years!

But was any form of safety testing conducted during ancient times? The accounts are somewhat slim, but it appears chemists of the time were required to perform some measure of quality control after preparing a medicine. What exactly those standards were is less certain, but it's probably safe to say that some cautionary steps were being taken without overstating the case for drug safety testing. For example, in the "Recipe for Transforming an Old Man into a Youth," the language clearly states that when the water from the washing of the compound is not bitter to the taste any longer, then the formula is correct. We now know that bitterness is the cyanide coming out of the mixture.

The evidence shows, however, that the ingredients used to mix a formula were almost always measured by volume as opposed to weight. Measuring by weight is by far the more accurate standard, particularly when dealing with a substance such as a grain or herbs that can vary in water content or unevenly hide pockets of air when packed in a measuring cup. In other words, there can be varying density so it's important to be precise. But the most common method of formulating a compound prescription was to specify fractional proportions for each component—1/2 flour, 1/4 sugar, 1/4 baking powder, for example—regardless of the overall quantity of the drug being made. How the ancient Egyptians then calculated the proper dosage for a patient is less understood, but it would have been vital in the case of ingredients that contain highly toxic substances.

I continued my research and was increasingly impressed with the medical practices and techniques of ancient Egyptians, but after three years of buying books, searching the Internet and making inquiries of a few Egyptologists associated with ARCE (American Research Center of Egypt), I was no closer to deciphering the mysterious hemayet fruit than when I first found the recipe in Breasted's book. I was becoming discouraged. And even more aggravating was knowing the possibility of what this elixir could mean for skin care today:

> *"Anoint a man therewith. It is a remover of [wrinkles] from the head. When the flesh is smeared therewith, it becomes a beautifier of the skin, a remover of [blemishes], of all [disfigurements], of all signs of age, of all [weaknesses] that are in the flesh."*

The passage seemed so meaningful and attainable, as the scribe noted: proven good *a million times*. A million times. This line took hold inside me and continued to weave through my thoughts in the weeks and months ahead.

But what was hemayet?

"It suggests that ancient Egyptians had performed a huge clinical trial," I said to Joni one evening. "They must've tried this chemical process hundreds of times to get it right."

"It certainly seems so. And if it works, imagine! Every woman I know would want a wrinkle remover, despite it being only a 'face paste,'" Joni said, pausing to smile. Knowing what we know today about the obsession of women around the world with antiaging and skin care, she was amused as I was by Breasted's comment that the recipe was "...simply a face paste, which will remove wrinkles."

It was becoming apparent that my steps had to take me closer to the birthplace of this ancient formula. I knew it, and my family was coming to the same realization as well.

One autumn evening in 2004, Joni and I were sitting in the family room when I showed her a promo from ARCE I'd received via email for a trip to Egypt. There were amazing images of the Sphinx, the great pyramids of Giza, the Karnak Temple and Valley of the Kings. It was an ARCE-sponsored excursion, organized by Janice Brannon of Seven Wonders Travel in Chicago, and Mohamed Nazmy, owner of Quest Travel in Cairo, both of whom would accompany us.

"What do you think?" I asked her. We were on the sofa, and the boys were in their pajamas, transfixed by a television program. It was a three-week trip, scheduled for February, the middle of the school year for the boys. Austin was in the sixth grade at the time, and Spencer was in third grade.

"I think it's a great opportunity," she said.

"Should we take the boys?" I asked quietly. "Would their teachers be okay with this?" I knew she'd hate to pull them out of school as much as I would, but this trip would provide an education far beyond reading about Egyptian history in a textbook. And we both enjoyed travel and wanted Austin and Spencer to be ambassadors of the world, learning about other cultures, languages, history and food. It seemed an opportunity of a lifetime, and surely someone in Egypt had to know the meaning of the word hemayet.

"Believe me, I want to go and think we should explore the idea, but..."

I looked at her. "But what?"

"Is it safe? The political climate right now is so uncertain...and I'm a little apprehensive about taking the boys there."

Her concerns were valid. "I'll check into that before booking the trip."

"Trip? Are we going on a trip? Where are we going?" Austin asked.

"Egypt...*possibly*," Joni said. "But you'd have to miss school for three weeks."

"What? We're going to Egypt...for three weeks?" Spencer asked. "Three *whole* weeks? I can miss school!"

"Me, too!" Austin chimed in.

Suddenly, the program they'd been watching on TV no longer captivated the boys. They were good students and enjoyed school for the most part, despite the usual grumbles about homework. But they were both all bright-eyed enthusiasm and more than happy to sacrifice nearly a month of school to help solve the mystery of hemayet.

I glanced at Joni and we shared a smile over their eager enthusiasm. *"We'll see,"* she said. "Let me speak with your teachers."

Chapter VII

Cairo, Egypt
February 2005

M y first impression of Egypt was sand. And lots of it. Not so much a surprise, looking out the window of the jet as we flew over the seaport city of Alexandria on the Mediterranean coast. As we prepared to land at Cairo International Airport, the pilot followed a route more or less along the Nile. Beyond the delta and the riverbanks, I saw nothing but sand on the horizon, then a speck of geometric outcropping growing up from the desert—Cairo.

From the plane, Cairo was a sprawling city that looked mostly beige, blending in with the sand surrounding it. The buildings were not exceptionally tall, and most were almost uniformly constructed with some sort of concrete-type of block. But once you step onto terra firma, "beige" is hardly the word to describe this fascinating and densely populated region of the world, occupied by more than 16 million people.

The city is bustling by day, but seems to come even more alive at night with the festive atmosphere of the souks, marketplaces. Several of these marketplaces dot Cairo, and the great bazaar downtown is crowded at all hours. At sundown, a light comes on at every corner, illuminating thousands of vendors selling everything from gold coins and flat breads to spices. Bargaining is routine, as it has been there for centuries, and no one expects you to accept the starting price. People jostle shoulder-to-shoulder, in many locations, along the narrow streets that traverse the souk.

We felt safe in Cairo, but with our two young sons in tow, Joni and I kept a watchful eye on them. It was difficult sometimes to move through throngs of people in the marketplaces as a family. Joni was easy to spot with her blonde hair, but the boys blended in with their dark hair and olive skin.

One afternoon we had just finished lunch with the ARCE group at a restaurant in the middle of Khan el-Khalili, a major souk of Cairo. As we left the restaurant, we melted into a throng of people walking by in many different directions because of the small interwoven streets, alleys, tiny passages, and a multitude of stores. As locals streamed passed us, carrying goods on their heads, I made my usual, rote head count when I realized Spencer was missing.

"Where's Spencer?" I asked Joni, trying to sound calm, but my mouth had gone dry and I could feel my heart pounding.

"He's right here…," she said, turning and glancing around. Austin was with us, but no Spencer.
"*Ohmygod!* Austin, where's Spence?" She asked, her voice growing increasingly frantic.

"I don't know," he said, looking behind him.
Spencer had wandered off in a split-second, and as we surveyed the area around us, he was nowhere to be seen. It was noisy—people were talking and shouting in Arabic and English, vendors announcing their wares and offering samples. It was a situation where a small boy could easily disappear. The three of us began searching for him.

Fortunately, we found Spencer within minutes. He had wandered ahead of our group in the souk, thinking we were only a few steps behind.

"*Spencer!*" Joni called, half-walking, half-running toward him. He had his back to us, moving through the crowd, happy-go-lucky and completely oblivious. "*Spencer!*"

He turned, caught off guard by his mother's panicked tone. "Don't ever do that again!"

"Do what?" He asked, looking up at his mum.

"Run off ahead! We stay together," she said.

"But I was right here, and you were right behind me," he said, wide-eyed and astonished that we were upset.

Joni and I shared a glance, relief washing over both of us. The heart-pounding experience was brief, but we had a stern talk with the boys about staying together at all times and not wandering off.

Our first excursion was to the Great Pyramids of Giza. The pyramids rise up from a huge plateau that is partially hidden because of the city's buildings. After driving past a modern-day mecca of Burger King and Kentucky Fried Chicken franchise eateries, we suddenly had a view of ancient times in all its majestic glory. Its beauty made me pause and catch my breath. Laid out in a trio, the Great Pyramid of Khufu is the first one you see. Its thousands of stone blocks hewn by hand, had been placed in perfect geometric form to reach the sky. Though it looked to be a simple building at first glance, its sheer size and the fact it had survived more than 4,000 years is humbling.

The Land of the Pyramids is hot with sustained temperatures in excess of 90 degrees Fahrenheit during the summer months. But during the winter months, known as "digging season," the temperature is more moderate and it's the only time of year when it's comfortable for archaeologists and their teams to safely conduct excavations. Winter is naturally the peak season for tourists as well.

The Great Pyramids on the Giza Plateau were one of our first stops on the tour led by Dr. Gerry Scott, the director of ARCE's Cairo office. There were about 10 Egyptologists from all over the United States on the excursion, including one professor from Hawaii, which made for a highly educational tour. My family and I appeared to be the only non-Egyptologist or historians on this excursion. We learned from them as we traveled in the bus from site to site. As we visited the various tombs, we enjoyed listening to the professors debate the various possible meanings of a single hieroglyph. Though it was entertaining and fun to watch, it was also anxiety-provoking as it made me realize that interpreting this ancient language—and my mystery word—might be even more difficult than I'd expected. Naturally, I asked every Egyptologist on this excursion, including Dr. Scott, if they'd heard of hemayet fruit. The answer was always the same: No.

Ancient Egyptian Skin Care

If necessity is the mother of invention, then Egypt's relentlessly dry climate no doubt prompted ancient Egyptians to develop all manner of skin care preparations. Before currency came into existence, oil for external use was often included among other goods in a worker's wages.

With hygiene and skin care being of vital importance to them ancient Egyptians in life, the same necessities naturally followed them into the afterlife as well. Many unguent and perfume containers, carved from alabaster and other precious stones, have been recovered from King Tutankhamen's tomb.

Strike one.

But we did learn a great deal in the company of these entertaining Egyptologists, occasionally bordering on the bizarre. One particularly intriguing tale had to do with what is known in Egypt today as "batted villages." These places supposedly get their name from the ancient practice of applying bat's blood to an infant's underarm and groin area, which the locals say prevents hair from ever growing in those areas. That's the trouble with medical folklore, however. Some of it works; some of it is pure malarkey. (Applying butter to soothe a burn, for example, only intensifies the injury.) Sorting out fact from fiction takes scientific research, which can't always easily (or ethically) be conducted. But I found the story fascinating, as there may indeed be some complex molecules in the blood of bats that could inhibit hair follicle growth in human babies, but it's the perfect example of an experiment that would be nearly impossible to conduct.

The Great Pyramid of Khufu is the largest of the three pyramids on the Giza Plateau as well as the last surviving monument of the Seven Wonders of the Ancient World. Constructed as the tomb for Pharaoh Khufu, it took between 10 to 20 years to construct, and was completed around 2521 B.C., preceding the death of Khufu by about seven years.

The best way I can describe the Great Pyramid of Khufu is to say that it blocks out the sun. And each block at the lowest level of the 450-foot-high structure is much larger than I am (about six feet) in height, width and depth, but the blocks get smaller in size from the ground level, looking up. It must've been quite a view from the top of this pyramid, which could be accessed by tourists willing to make the climb up until the 1980s or so. The climb up was not easily made, however, and more than a few people fell to their deaths. These giant blocks of limestone were once overlaid with casing stones of an even finer white, smoothly polished limestone, and there is some speculation over whether a gold-covered capstone may have adorned the pinnacle of the pyramid. Plans to place a new one atop the structure to celebrate the millennium in 2000 never came to fruition. Regardless, it was easy to imagine how magnificent this towering monument must've looked to people in ancient times as they approached it from a distance, the sun reflecting off its limestone siding.

The size of the three Great Pyramids relative to one another is deceptive. The two pyramids of the pharaohs Khafre and Menkaure were built after Khufu's and are shorter in height—Khafre's pyramid by 10 feet or so, and Menkaure's pyramid is noticeably smaller by about 200 feet. We toured the areas surrounding all three pyramids where hundreds of tombs for the queens, children and other relatives of the pharaohs are located. Behind Khufu's pyramid, we saw a museum built specifically to house a magnificent ancient Egyptian boat crafted of cedar wood, which was discovered in a dismantled state during an excavation in the 1950s. Re-assembled by the 1960s, the "Solar Boat" is a fully functioning, 140-foot-long vessel with multiple sets of oars. The boat's ropes are still in remarkably good condition, showing how the hemp used to make them was twisted together in an intricate pattern of interweaving. It's uncertain, however, whether the boat was ever used to transport the king's body to its final resting place, or simply symbolic of his journey to the afterlife.

Later that day, Joni, the boys and I ventured inside the Great Pyramid of Khufu, and it proved to be an adventure that challenged all the senses. The 1908 edition of Baedeker's *Egypt* cautions: "Travelers who are in the slightest degree predisposed to apoplectic or fainting fits, and ladies traveling alone, should not attempt to penetrate into these stifling recesses."

The Science Behind the Early Days of Eyeliner

Eyeliner was first used in Ancient Egypt and Mesopotamia as a dark black line around the eyes, according to an online search. As early as 10,000 B.C., Egyptians wore various cosmetics not only for aesthetics, but to protect the skin from the desert sun.

Kohl, the prominent eyeliner featured in nearly all the art depicting men and women—and animals as well—is an intriguing example of where cosmetics and medicine intersect. Applied around the eyes with a small stick, the dark substance reduced the intense reflection of the sun off the sand and repelled flies, a common source of eye disease that often led to blindness, according to an article on the ophthalmology of the pharaohs in the *British Journal of Medicine* in 1909.

In 1922, Tutankhamen's tomb was discovered by the English Egyptologist Howard Carter. It was at this time the use of eyeliner was introduced to the modern world. Meanwhile, there were many changes in women's fashion and women began applying makeup more liberally.

Today, eyeliner is commonly used in women's daily makeup routines to define and cosmetically enhance the eyes, but it still serves a medicinal purpose as well. American football players and baseball players still wear streaks of "eye black" to cut the glare of the sun by day and stadium lights at night. (Fortunately, these athletes don't have to worry about using eye black to keep pests out of their eyes as they did in ancient Egypt.)

Ancient Egyptians in the earliest dynasties wore kohl compounded from powdered antimony with green malachite on their lower lids and black galena on their upper lids. It appears less frequently in their art, but statues depicting this green and black combination can be seen today in the Egyptian Museum in Cairo. Modern technology has allowed us to dig deeper into the composition of some of these ancient formulas, revealing a level of sophistication previously undetected.

Recent analyses of minute amounts of the contents of 49 kohl containers in the Louvre dating from 2000 to 1200 BC have revealed that the ancient Egyptians were able to use "wet chemistry." The kohl was found to contain substantial amounts of laurionite and phosgenite, which on their own have the appearance of white powder. These don't occur naturally in quantity and must have been produced synthetically by a laborious process of admixing rock salt and/or natron in water and repetitive filtration.

Coincidentally, wet chemistry was one of the few aspects I could readily decipher about the three-millennia-old mystery I had on my hands. But it was just the start.

Apart from the outdated language and silly part about women traveling alone, it's true. The experience requires an adventurous spirit, and the claustrophobic should not apply!

The four of us dared to venture inside, we took the front staircase climbing a short way up the pyramid's front wall to reach a small entrance. Inside the first foyer, there was a room for all but the tallest people to stand. From there, the journey intensified. We climbed into a steep, roughly four-by-four-foot stone shaft with wooden slats on the ground for traction and handrails along the walls. Ascending in a cramped posture for about 200 feet, we reached the Queen's Chamber, which was stabilized by massive granite blocks placed against one another to support the ceiling. We climbed a series of ladders through the Grand Gallery before reaching the King's Chamber where Khufu was supposedly buried.

I say "supposedly" because the great pharaohs may not have been entombed in the obvious chambers within their pyramids, and the final resting place of Khufu is unknown. The walls inside his tomb were undecorated and a single granite sarcophagus (stone coffin) stood empty on the floor. The lid has disappeared, and visitors are allowed to climb inside the coffin for a photo opportunity. We all took turns climbing into the pharaoh's sarcophagus.

Though Joni and the boys have an adventurous spirit when it comes to exploring new places, entering the pyramid was a bit of a spine-chilling experience. The blocks inside are fitted so perfectly together without mortar that not even a credit card can be slid between them. The shafts and tombs were dimly lit, silent and the air was dank and still. Only a few groups of people are allowed to enter each day. During our tour, visitors in another group were chanting mystical-sounding songs that contributed to the atmosphere. It felt as if an ancient procession was about to pass by, taking us thousands of years into the past. It reminded me of the movie, "The Mummy," when the heroine Evelyn O'Connell is transported back in time to fight her sister.

After the tour and back in the open air, I continued my search for the elusive hemayet fruit. Earlier in the day, I had asked various traveling companions if they knew about any medicines or cosmetics ancient

Cleanliness & Godliness

We can trace so much of what we currently use to heal and enhance our bodies with directly to ancient Egypt. But what beliefs drove this civilization's intense pursuit of physical well-being and beauty?

Cleanliness was indeed next to godliness, as ancient Egyptian priests were required to rid themselves of nearly all body hair before performing rituals. But this religious practice had a highly practical side as well. Hair removal by people in all walks of ancient life was common and necessitated by the prevalence of lice. As a result, women and men commonly shaved their heads bald, which served a dual purpose. The wigs discouraged lice and made it easier to endure the hot Egyptian climate. Wigs and hairpieces were most often made of human hair and styled with beeswax and were almost always worn by women when in public. Much like today, it appears baldness was considered a look more aesthetically suited to men.

As for the lice, these itchy little insects apparently continue to annoy the Egyptians in the afterlife as well. The carcasses of the same type of head lice that occasionally pester today's schoolchildren have been found on mummies thousands of years later.

Egyptians used to improve their appearance. One Egyptologist suggested a substance known as "royal amber," a nourishing, perfumed oil applied to the skin.

Not knowing where to start, I asked Mohamed Nazmy, who was leading out tour and lives in Cairo. He suggested I visit the Egyptian Perfume Palace, which was not far from the Great Pyramids, so we stopped by on the way back to our hotel.

Though the entrance was simple and unassuming, the inside was strikingly different. The opulent shop was a magical apothecary. Dream-like, it gleamed with bottles and bottles of perfumes and oils in various colors and fascinating shapes, on dozens of glass shelves. The lights refracted off the glass to create a diamond glitter.

The shopping experience in Egypt is remarkably different than in the United States. The ritual in nearly every shop is to sit down with the proprietor over mint tea, talk about your family, what you do for a living and so on. I found it a pleasurable and civilized way of

King Tut's Tomb

When archeologist Howard Carter famously recalled his first glimpse inside King Tutankhamen's tomb in 1922, he wrote of seeing: "…strange animals, statues, and gold—everywhere the glint of gold." He was referring, of course, to the magnificent gold-leafed funereal furniture adorned with carvings of animal gods stacked in the tomb's antechamber. But these were not only the sacred items within. The soul or "ka" of a person needed food to sustain him in his journey to the afterlife, so the boy king was supplied with a final feast fit for a pharaoh.

What was included in the royal spread? There were more than a dozen loaves of bread, two jars of honey, 27 jugs of wine, more than 45 boxes of prepared oxen parts, trussed geese, pigeons, duck and squab. Tucked into baskets and other containers were almonds, black cumin, thyme, dom-palm fruit, grewia fruit, cocculus fruit, persea fruit, and seeds of chervil, coriander, safflower, sesame and watermelon. A small model granary contained samples of barley, chickpeas, emmer wheat, garden peas, lentils and fenugreek and provided a veritable botanist's wonderland of plant specimens known to the ancients.

doing business. Egyptian proprietors want to get to know you before negotiating the sale. What's more, bargaining over price is a way of life there, so your final cost largely depends on how well you interact socially. If a person behaves politely, the proprietor is more inclined to bring out his finest products and offer better deals.

The mustached owner of the Egyptian Perfume Palace sat us down on soft stools, offering us tea and then explained the history of the store and the various oils and perfumes he offered.

"How is it possible to have all these oils in one store?" Joni asked the proprietor.

I was curious, too. I wondered which vials contained a substance that really worked and which were only for show.

His explanation was somewhat confusing, and the gray area became even grayer. There were thousands of years of history within this store, but also thousands of years of deceiving the public.

Pyramid Builders

Old Hollywood movies often depict pyramids being built by half-starved slaves, being whipped into submission by cruel foremen, but the portrayal is wide off the mark.

American Egyptologist Mark Lehner, Ph.D., internationally-known and highly respected by his peers, has devoted more than 30 years of research to the Giza Plateau since enrolling at the American University of Cairo in 1973. He directed the Sphinx and Isis Temple projects for five years, and since 1983 has led the Giza Plateau Mapping Project to research the topography of the region and study the function of its tombs and monuments. At the time, working with Zahi Hawass, Ph.D., Egypt's secretary general of the Supreme Council of Antiquities, the two have overseen the last 15 years of excavations of a village that is thought to have housed up to 20,000 pyramid workers. And what their research has brought to life debunks some popular myths.

The work of Lehner and Hawass in this century has given rise to the idea that the pyramids were not built by slaves, but by regular, able-bodied citizens drafted to perform their service before the pharaoh for several months of the year. Depending on their skills, workers appear to have rotated through their specific jobs as bakers, builders, doctors, scribes and so on before returning to their home villages to carry on with their lives. During the trip to Egypt, our ARCE group toured the remains of buildings where the workers slept, prepared meals and completed other tasks necessary to keep a workforce healthy and productive, located within the shadows of the pyramids It was a vast expanse, easily equal to three or four football fields. In fact, the eastern section wasn't discovered for many years because a soccer field lay in the area. As the archaeologists received permission for their digs, more and more of the city was exposed.

Nevertheless, he sold some truly useful unguents. The proprietor showed us various oils, and I must've followed the customary Arabic politeness reasonably well, as I was shown some of the more unusual oils in addition to the royal amber, which was the color of dark rum when held up to the light.

The Egyptologist who told me about the royal amber said it might contain an extract from the salivary glands of an antelope, a form of musk. I lifted the cap from the oil and sniffed the contents. The scent of the royal amber somewhat surprised me because it had a sweet smell, almost like that of a floral perfume with a hint of baby powder.

"Does this by chance contain hemayet fruit?" I asked him.

"No," he said, shaking his head.

"Do you know who might know what hemayet is?"

He shook his head. "I've never heard of it."

I purchased three bottles of oils—sandalwood, almond and the royal amber. Meanwhile, the proprietor inquired among his employees about the meaning of hemayet fruit, but the answer was always the same: No.

Strike two. The meaning of hemayet remained a mystery.

Chapter VIII

Aswan, Egypt
January—February 2006

awn breaks over the Nile, its deep sapphire waters brightening to lapis blue with the growing light of the morning. After two weeks in Egypt, the beauty of this land was still not lost on any of us. I'd purchased a beige Four Season ball cap at the hotel for the trip and was wearing it as we boarded a cruise ship in the early morning. We were about to embark on a four-day, round-trip journey south to Aswan, stopping at ancient sites along the way, including the Valley of the Kings.

From the moment we set foot in the Valley of the Kings, it became one of my favorite places on Earth. Carved into sedimentary rock at the base of the Theban Hills, the area served as a necropolis for some of ancient Egypt's most powerful rulers from about 1500 to 1075 BC. Amenhotep II, several bearing the name Thutmose, Ramses the Great (II) and the Ramses kings to follow as well as Tutankhamen were all buried here. More than 60 tombs have been identified, including some for the queens and children of the pharaohs.

Robbed empty of all their treasures long ago, only Tutankhamen's tomb was found relatively intact during its celebrated discovery in 1922. What remains for people to see inside these burial sites are the spectacular wall drawings. Unlike the barren walls of the King's Chamber in Khufu's Great Pyramid, many tombs in the Valley of the Kings have been decorated with scenes from ancient Egypt's "Netherworld Books" and other symbology that was thought to help the deceased travel through the afterlife. But it was the tomb of Pharaoh Seti I that I found the most beautiful of all.

Stretching almost 400 feet, it's the longest of all the tombs in the Valley of the Kings, discovered in 1817 by Giovanni Belzoni, the same Italian adventurer who provided the account of finding papyri wrapped inside mummy casings. The walls include magnificent representations of the gods Hathor, Thoth, Sekhmet, Osiris and others, but the artwork I found most breathtaking was on the ceiling of the burial chamber. Painted with astronomical symbols, it depicts the constellations on a vivid blue night sky, along with a timeline of calendar units. The god, Isis, her wings stretched across the entire heavens, keeps the sky and earth safe. The colors were so vibrant it felt as if the artist might have just finished and stepped into the chamber next door.

The tomb of Seti I is also the deepest and has a stairwell that descends to meet an intersection with two paths. We were directed to the right hand passageway that spans a 20-foot long bridge passing over what appeared to be a miniature moat, a small pool of still, dark groundwater about 15 feet below. Curious and caught up in the moment, I bent over the railing to have a closer look, and the hat I had purchased that very morning, fell off my head and slid into the water, barely making a sound before I could make a grab for it.

The four of us stood frozen, looking at the hat. We knew there was no way to retrieve it. Then the humor of it all struck a chord and we laughed.

In the excitement of the day, I nearly forgot the area around Thebes was where The *Edwin Smith Papyrus* reportedly had been found. As our ship sailed south again toward Kom Ombo, I wondered whether any of our next steps would lead me to something or someone who would as last reveal the identity of the puzzling hemayet fruit. Kom Ombo didn't turn out to be the place where that happened, but it was full of marvels nonetheless.

In the Temple of Kom Ombo, we saw depictions of ancient Egyptian surgical instruments and the Eye of Horus "Rx" symbols. (See "The Rx Symbol" on page 60) We were also shown the priests' chambers with hidden vents that connected to various chambers for worshipers. This clever arrangement was thought to allow the priests to speak without being seen, presumably convincing the person praying that he or she was communicating directly with the gods. Next on the itinerary, and not too far from the Temple of Kom Ombo, we visited one of several known "Nilometers," which are round, well-like structures dug deep into the ground at the river's edge. Down a set of stairs inside the 15-foot-wide chasm was a floatation device with water-level marks on the walls. In a clever fashion, the Nilometer enabled engineers to gauge the probable height of the Nile each season, so they could make preparations in case of a disastrously low or high inundation of rain. Thus, the pharaoh could tell his subjects whether the Nile was going to flood, and if so, to store more grain. These predictions would then come to pass, making the pharaoh look like a god.

Our trip to Egypt was nearly at an end when our ship next docked in the waters near Aswan High Dam. At this site, imperiled monuments such as the temple of Ramses the II, which was carved into the side of a mountain, were moved to higher ground at Abu Simbel during the 1960s. I continued exploring the area and inquiring with experts about the translation of hemayet and though I was still no closer to discovering its meaning, I remained hopeful.

Once we returned to Cairo, I managed to buy a book or two before leaving that I hoped might contain information about hemayet fruit. One was a fairly large hardcover book, *Egyptian Luxuries: Fragrance, Aromatherapy and Cosmetics in Pharaonic Times,* by Danish Egyptologist Lise Manniche. The deep blue paper cover, reminded me of the color of the ceiling inside Seti's tomb. I smiled at the thought, hoping this might be a good omen. The book was too awkward to stash in a carry-on bag, so I packed it in my suitcase.

At the airport the next morning, we settled into our seats, and I glanced out the window for one last look. Egypt was a sea of sand, yet it had an enchanted drawing power. I wasn't prepared for the sadness I felt upon leaving.

"I'm not ready to leave," Joni said.

"We'll be back," I said, reaching out and squeezing her hand. I knew exactly how she felt. I had the sense I was leaving home, as if this is where it all started. Where we all, in a way, were born. Other civilizations existed before ancient Egypt, but none before or since has flourished as long or left such an impressive blueprint for future societies to follow. From astronomy to surgery, cosmetics to construction and everything in between, there are arguably few disciplines where this civilization has not left its mark. I glanced at the boys. What school child isn't captivated when first learning about ancient Egypt? The pyramids, the mummies, the discovery of King Tutankhamen's tomb and the legend of Cleopatra. I instinctively knew this trip was indelibly imprinted on both Austin and Spencer; they'd remember their time here for years to come. Our enduring fascination with these incredible people is somehow encoded deep within our DNA.

There was the usual cacophony of sound as the plane prepared for takeoff—the roar of the engines, the low hum of passengers talking and the click of seat belts. As the plane lifted off, I turned my attention to the land below. Though I hadn't yet found the answer to my personal quest, I knew with certainty we'd be back. I could see the pyramids in the distance and marveled once again at their construction. It made sense to me that people this exacting with their construction of monuments were likely to be no less precise when it came to the practice of their medicine. I was more determined than ever to solve the 3,600-year-old medical mystery of hemayet, and I hoped the book I'd packed in my luggage held the answer.

The mystery and intrigue of ancient Egypt and its culture continued to consume me in the weeks that followed our arrival home. At night I dreamed of the places we visited, and during the day, I considered the places I wanted to visit when we returned someday. The place was magnetic. Fortunately, my family felt the same way and we were already talking about our next trip.

I began reading Manniche's book, *Egyptian Luxuries: Fragrance, Aromatherapy, and Cosmetics in Pharaonic Times*. The book was beautifully illustrated with photographs of elegantly carved vessels that ancient Egyptians used to store their cosmetic "luxuries." It also included recipes and details about some of the most common ingredients in many ancient formulas. Much like Breasted's book, I read it at night in my living room. One evening, I saw a sentence on page 118 of Manniche's book that startled me: *"In the medical texts, the word hemayt is taken to mean fenugreek*."

The word was missing the "e," in hemayet, but it had to be the same. In her book, Manniche explains that "hemayt" was the "sole ingredient" in a formula that was remarkably similar to "Recipe for Transforming an Old Man into a Youth." After six years, it appeared my search for the mysterious hemayet fruit had finally come to an end. I closed the book, feeling elated that night.

The Rx Symbol

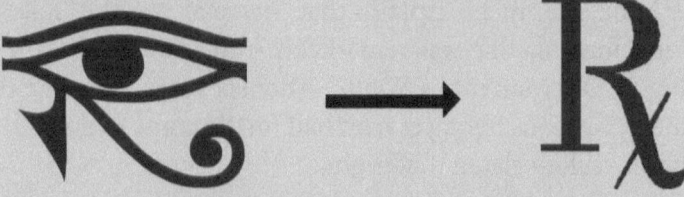

Rx is the internationally-recognized symbol for prescription medicine, and this curious combination of letters directs the pharmacist to dispense medication per a physician's orders. But where did Rx come from?

Theories vary to some degree, but intriguing evidence points to the symbol's origin in Egypt more than 5,000 years ago with the legend of the Eye of Horus. According to ancient Egyptian mythology, the falcon-headed god, Horus, lost his vision as a child, following an attack by the evil god, Seth. Magically restored by the god, Thoth, the Eye of Horus became a symbol of healing and godly protection still worn as an amulet by many Egyptians today.

During the Middle Ages, the Eye of Horus reappeared in modified form, which physicians wrote on their prescriptions to invoke the assistance of Jupiter. Over time, this sign for Jupiter evolved into the letters "Rx."

I'd been practicing medicine for more than 20 years before learning of this incredible connection and seeing for myself more evidence to support it at Kom Ombo, a Ptolemy-era temple, along the east bank of the Nile about 400 miles south of Cairo. Inscribed on the walls within, along with rare images of what appear to be early surgical instruments, are numerous representations of the Eye of Horus. When drawn quickly, the sloping line that descends from the middle of the eye, visually creating the open portion of the capital letter "R," can get crossed at its tail, forming an "x"—making an Rx.

I learned more information in the temple that day that supports this theory and returns us to what we know about kohl. Because it was also a medicine that protected the eye from disease, the kohl-lined eye became the symbol for treatment, transforming into the symbol for prescription medicine later in history.

Plants of the Pharaohs

The Egyptians experimented extensively with plant matter in medicines, documenting many of their recipes in the voluminous *Ebers Papyrus*. Do we know today if any of these plants work as the ancient Egyptians intended? Yes, it's widely known that figs, for example, have laxative properties, but it's doubtful there's anything in lettuce that initiates hair growth.

In addition, unprocessed castor beans contain ricin, a highly toxic poison, and unprocessed aloe contains the compound aloin, which can cause profound diarrhea, and provoke contractions in pregnant women. (As they say, the dose makes the poison, so don't make this at home.) Here are some of the plants identified in those ancient formulas and how they were used.

Aloe vera: Juice of plant used as a laxative.

Frankincense (*boswellia*): An anti-inflammatory.

Celery* *(apium graveolones)*: Used as a digestive aid, and as a poultice applied to burns.

Cumin *(cumin cyminum)*: Used as a digestive aid.

Dill *(anethum graveolens)*: seeds used in poultices.

Fig tree *(ficus carica)*: fruit used in laxative recipes.

Garlic* *(allium sativum)*: used in ointment and thought to be regularly served to pyramid builders along with onions and radishes.

Grapes *(vitis vinifera)*: wine, raisins and grapes used in myriad recipes, consumed widely as food.

Onion *(allium cepa)*: juice used as diuretic and expectorant, vegetable widely consumed as food.

Parsley *(apium petroselinum)*: used as diuretic.

Pomegranate *(punica granatum)*: juice used as digestive aid.

Sycamore fig* *(ficus sycomorus)*: fruit used in laxative recipes.

Wormwood *(artemisia absinthium)*: used to expel worms (bitter leaves of plant flavor, today's vermouth and absinthe.

*Also found in King Tutankhamen's tomb
Source: An Ancient Egyptian Herbal

Part II: Understanding the Structure of Skin

"Glowing skin is a result of proper skin care. It means you can wear less make up and let skin shine through."

—*Michael Coulombe*

Chapter IX

The Fabric of Skin

Your skin is a living, breathing fabric, and collagen and elastin are the two primary elements that give it structure. The elastin makes skin elastic and springy, and the collagen makes it plump, giving it thickness and viscosity. These two materials create primary and secondary lines on the skin, creating a crisscross weave, much like fabric. When you take special care of a sweater, for example, it makes a difference and the fabric lasts longer and maintains its appearance, rather than sagging, making it appear old and worn. But the difference between fabric and skin is that skin has the ability to rebound and regenerate at any age because your skin is a living, breathing fabric.

Yes, even some people who are in their 70s have the ability to generate new collagen and elastin. I'm often asked about this in plastic surgery. Patients say my skin is thin and won't tolerate a chemical peel, for example. But in reality, their skin can tolerate a peel since the skin becomes thicker as the skin cells produce more elastin and collagen. A TCA (tri-chloroacetic acid) peel will produce about a 50- to 70-percent increase in skin elasticity, but we hope some effective skin care products might produce a 15- to 30-percent increase, which is great for your skin. Why is improving your skin visco-elasticity important? Because improving visco-elasticity improves almost all aspects of your skin: wrinkles, lines, sagging, pores, tone, texture, luster and (indirectly) hydration.

The aging process

As we age, skin undergoes a great deal of damage. Sunlight, smoking, wind, pollution, acne and self-inflicted trauma (from scrubbing or using overly-abrasive cleansers), all break skin down. Collagen decreases by about age 40, and a once taut jaw line typically becomes less defined. At the same time, that decreased collagen production causes skin to sag and changes to collagen quality also become evident and shows up with the appearance of fine lines and wrinkles. The horizontal secondary fibers start to diminish. Skin gets stringy, loose and slack. Microscopic pictures show this. While a facelift can dramatically improve your appearance, making you look years younger, the skin—or fabric—doesn't change. But the good news is that whether you're 25- or 55- years old or beyond, there are steps you can take each day to achieve or maintain beautiful skin.

Anatomy of a Wrinkle

Did you know the loss of elasticity starts to occur 20 years before wrinkles appear? Results of a Japanese study conducted in 2011 by Skin Research and Technology, proves that how you take care of your skin today, will show up two decades later. That means that if you are 20 years old, your skin elasticity is starting to diminish, so that when you are 40, you will start to see fine lines and wrinkles. The moral of the story is to start taking care of your skin at an early age.

When you spend extended periods of time in the sun, you need to wear sunscreen and sunglasses and a hat with a brim that will offer added protection to your face, neck and chest area. If you're a parent, remind your children to do the same. The habits you develop at 16, for example, will show up on your skin at age 36 and beyond.

Skin Facts

Skin is the largest organ of your body, and it's an amazing one. Skin is the first defense against infection and is under-recognized as to its importance of the immune system. Having healthy intact skin is important for overall health. It can stretch and expand and has the ability to contort and change shape, which allows surgeons such as myself to perform plastic surgery. In fact, the word plastic comes from the Greek word plastikos—to change shape. Having operated on more than 12,000 patients in my career, I've learned that there are more similarities than differences in skin. There's thick skin and thin skin, oily skin and dry skin, but we all have similar melanocyte, collagen and elastin.

Here are a few facts about your skin:

• **Your best skin is on your face.** The reason is that facial skin has more skin-forming units, which means more specialized cells to regenerate new skin. The worst skin on your body (in case you're wondering) is the skin on your shins.

• **The skin under your eyes and around the eyelids is the thinnest skin.** At 0.5mm thick, it is one of the first areas to show signs of aging such as "crow's feet" and wrinkles.

• **The skin of your throat is often the most flawless skin on your face and neck.** This is something I've always noticed before beginning surgery. This is because the area just under the jaw has been shadowed from sunlight and is silky smooth and soft, and not discolored. The skin on the sides of the neck, or below the hyoid bond, show the ravages of cosmic rays.

Chapter X

Skin Care Basics

I f you're looking to stall the hands of time on your biological clock and slow down the visible signs of aging, the good news is that food and lifestyle choices can minimize the impact. Taking care of your skin should be an essential part of your health care regimen, and it should be simple. Here are some basic strategies for taking care of you skin.

Cleansing your skin

Washing your face is a simple and important step to healthy skin. A simple soap such as Neutrogena, Dove or Cetaphil cleanser works well. I've researched an extensive list of chemicals used in makeup, and some of these chemicals can cause skin irritation or sensitivity, not to mention blocked pores. I recommend washing your face as soon as possible after wearing makeup—as soon as you get home from work is ideal, but certainly before going to bed.

I'm not opposed to the facial motorized brushes, as I realize they may allow mechanical removal of some dead skin. Using an exfoliating agent to chemically remove dead skin or *stratum corneum* is more effective by encouraging a rapid turnover of new skin cells or *fibroblasts*. A good exfoliating agent can remove 5 million dead cells from your face in a matter of minutes. The peel pads we created for the Stanley Jacobs skincare product line do just that.

Protect your skin from the sun

We've all been told to protect our skin from the sun, but sometimes we still shrug off the advice wanting to achieve a so-called healthy glow. In the past, general scientific knowledge was poor, regarding skin cancer and sun exposure, but today we know spending time in the sun increases the risk of skin cancer and aging. Many women (my wife and her girlfriends included) used tanning oils and reflecting surfaces to achieve a sun-kissed glow, especially in the '70s and '80s. But today we know using sunscreen is critically important—and scientifically proven—to slow down the aging process and prevent skin cancer.

There are three types of skin cancer. Basal cell carcinoma is the most common cancer in the world, and is fortunately the least aggressive, as it never metastasizes. Squamous cell carcinoma is more aggressive and can potentially metastasize. And melanoma is a cancer that can be deadly. Sunscreens are highly important to minimize the risk for all types of skin cancer, and both UVA and UVB rays need blocking. Be sure to apply it to your neck and décolletage (the area of the chest skin from the collar bones to the upper breast area). About 75 percent of women that I see for a facelift or chemical peel have sun damage in this area, which is as important to protect from the sun as your face and requires moisturizer, sun protection, and topical agents to increase skin elasticity.

Kick the habit

One simple and dramatic way to improve the quality of your skin is to stop smoking. Smokers have sallow, dry, rough-looking skin that lacks the vibrant plumpness of healthy skin of a non-smoker. Despite the products you're using on your skin, or other healthy habits you've developed (using sunscreen or exercising, for example), smoking will cancel out your efforts.

Why is smoking bad for your skin? It shrinks small blood vessels and as a result, these vessels can't deliver oxygen or nutrients to your skin, which reduces the hydration (or plumping) to our skin. Smoking reduces blood supply to your skin, for example, the same way a garden hose would only leak out a trickle of water if you stood on it. Smoking shuts off nicotinic receptors, which are like the tap or valve to your blood vessels. When you stop smoking, it improves blood flow, even after three months. This improved blood flow benefits all tissues, including your skin as well as your heart and lungs. So, get off the garden hose and stop smoking if you want beautiful skin. When long-time smokers seek more from a surgical procedure such as a facelift (or mini facelift), I generally recommend they completely stop smoking for up to six months before scheduling surgery.

Exercise for more than a healthy glow

We all know exercise is key to managing weight and improving cardiovascular health, but did you know it's also great for your skin? It is clear that athletes heal faster. Exercise increases blood flow to all tissues. Surgeons call that *tissue perfusion,* which is when oxygen and other beneficial blood products work their way into the tissues and enrich the cells. All surgeons know that improved blood supply leads to better wound healing and a better chance of surgical success. And it may surprise you to know that goes for any surgery patient—whether the procedure is a facelift or a hip replacement. Exercise increases the blood supply to all bodily tissues. There may even be new blood vessel growth with vigorous exercise. In fact, when someone suffers a heart attack, new vessels, known as *collaterals,* grow in other parts of the heart to compensate for the reduction in blood flow from the arteries that are plugged. This is known as *neovascularization.*

When I operate on a patient who exercises regularly, I know she will heal more quickly than a patient who does not. A 65-year-old woman who exercises routinely will heal more quickly than a 50-year-old woman who does not.

Best foods for good skin

A growing body of research points to inflammation as the common denominator to aging skin. According to an article published in the *Journal of Cosmetic Dermatology*: "Chronic inflammation appears strongly linked to many preventable and treatable skin diseases and conditions such as visible skin aging."

A healthful diet, rich in antioxidants, supports overall skin health. Generally, eating Mediterranean style, which is what most physicians are recommending these days to prevent cardiovascular disease and cancer, is not only good for preventing these diseases, it's good for your skin. That means eating foods with omega-3 fatty acids such as salmon, sardines and flaxseed, low-fat proteins such as fish, chicken and turkey as well as legumes. Load up on fresh vegetables and eat a salad every day—spinach, kale, romaine are loaded with

A Word About E-Cigarettes

Electronic cigarettes—known as e-cigarettes—are battery-operated devices that heat a liquid, turning it into a vapor that can be inhaled. Using e-cigarettes is often referred to as "vaping" and some the patients I've seen use them try to wean off cigarettes. There is no scientific evidence that lighting up with e-cigarettes is safe. My advice for healthy skin is to avoid them.

Most e-cigarettes have nicotine, which is the vaso-constrictor that reduces blood flow, just as cigarettes do. According to the Mayo Clinic, most experts agree that they're likely to cause fewer harmful effects than traditional cigarettes. However, researchers have found that some e-cigarettes have nicotine amounts that are quite different from what's described on the label as well as some flavoring agents and other additives have shown to be toxic. For more information, go to *www.mayoclinic.org.*

nutrients. When it comes to fruit, fresh berries and citrus fruits are packed with vitamin C and great for your skin. Replace the butter with olive oil. Limit alcoholic beverages and soft drinks. Remember, what's good for your heart and cancer prevention is also what's best for great skin.

As for supplements, if you're eating a well-balanced diet with plenty of fruits and vegetables, chances are your body is getting the nutrients it needs. The human body is a clever, multi-organ system that can extract all the nutrients needed from the food you eat. However, if you're on a restricted diet or have other health concerns, check with your doctor.

Get your beauty sleep

The average American logs in about six hours and 40 minutes of sleep a night, but seven to nine hours is ideal for muscle repair, memory consolidation and release of hormones that regulate growth and appetite, according to the National Sleep Foundation. Science also confirms that the repair of tissues occurs during sleep and skin builds collagen while you sleep, which plumps your skin. When you're tired, the blood flow to your skin is not adequate, which makes for a listless appearance. And when sleep is cut short, you may develop puffy eyelids. Get your beauty sleep. Aim for seven to nine hours of sleep each night for a rosy glow.

Drink plenty of water

Getting an adequate amount of water is good for overall health, but it's also important for you skin. Skin is an organ, and like other parts of your body, your skin is made up of cells. If your skin isn't getting a sufficient amount of water, the lack of hydration will make your skin dry and flaky. Dry skin has less resilience and is more prone to wrinkling. Most people are walking around dehydrated, especially in the winter. As a general rule, I recommend drinking at least eight glasses of water a day if you're minimally active. If you're an active person, or weigh more than 150 pounds, you may require three or more quarts of water a day.

Minimize salt

When sleep is cut short, you may develop puffy eyelids and some people are just more prone to them. Puffy eyelids can make you look tired. A simple solution is to minimize the amount of salt (sodium) in your diet. In fact, when my patients are contemplating lower eyelid surgery, I ask them to cut back on salt first. Keep your sodium intake to less than 2,000 milligrams a day. Avoid chips and salsa; and go easy with the saltshaker. And next time you order your favorite fast-food burger, check the nutrition

The Case *Against* Facial Exercise

Have you ever considered buying the latest facial gizmo to keep your skin supple and wrinkle-free? Don't bother. Electric currents (pizo electric) in machines applied to the face, simply make your muscles contract. The same as when you smile, grimace or frown continuously. They all cause wrinkles. (I was once asked to endorse such a product and studied the data produced by the company, but their data did not support the claims of their product. I suggested they change their study and try again, and never heard back from them.)

As for those neck exercises your mother taught you to do, don't waste your time. Neck exercises also cause wrinkles and promote what is more commonly known as "turkey neck." Bottom line: save your money and your time. These electric gizmos and neck exercises are not the equivalent of doing biceps curls to tone your body and develop muscle. "Crepey" skin can be improved with my serum. Sagging skin, however, is a natural part of aging and can only be corrected with a neck lift.

Making Sense of SPFs

If you've ever stood in a store trying to figure out which SPF is best for your skin, you'll be happy to know that in 2012 the Food and Drug Administration (FDA) finally stepped in to establish guidelines to protect consumers.

Under the new regulations, sunscreen products that protect against all types of sun-induced skin damage will be labeled "Broad Spectrum" and "SPF 15" (or higher on the front). Sunscreens should be at least 35 to 45 SPF. As for sunscreens that claim to have an SPF higher than 50, pure marketing hyperbole. They're no better. After SPF 50, the ingredients do not offer any further protection. SPF usually refers to UVB, but UVA is also cancer causing and aging for skin. Recently, the FDA reported that sunscreen should be at least UVA 10-plus to be effective. The UVA scale runs from zero to 12. Remember to apply sunscreen every two to three hours to be effective. For more information, go to *www.fda.gov.*

guidelines first. (One American favorite packs 4,000 milligrams of sodium.)

Reducing under eye dark circles
Dark under eye circles are a big concern for many of my patients. They are a result of pigmentation, fat bags, blood pooling (*venous stasis*) and puffiness. Getting enough sleep helps with blood pooling and fat bags can be corrected with *blepharoplasty* (lower eyelid lift) surgery. Pigmentation will improve with a topical such as mandelic acid, retinol and hydroquinone.

Jacobs family on camels

Jacobs family on Mt. Sinai

**Looking at hieroglyphs on
ancient columns at
Medinet Habu Temple, Luxor**

Dr. Stanley Jacobs exploring ruins at Karnak Temple, Luxor

Skin Cancer & Chemical Peels

A 1988 study by Mayo Clinic Proceedings reported that *tri-chlo-roacetic acid*—more commonly known as a TCA chemical peel—can reduce the chance of basal skin cancer by 50 percent. If you haven't been diligent about using sunscreen, a TCA chemical peel could dramatically reduce your chance of developing this type of skin cancer. See a facial plastic surgeon or dermatologist for a chemical peel.

Skin Care Basics for Men

When men seek a change in their appearance, the majority are more interested in correcting the signs of aging. Lucky for them, when men routinely shave, they're getting a daily microdermabrasion. The *stratum corneum* is shaved off and as a result, their skin gets less wrinkled and becomes thicker. And because their skin tends to be oilier than women's, moisturizers aren't as critical. However, because men tend to not be as diligent about using sunscreen and have shorter hair (or little to no hair), skin cancers are more prevalent on their lips, scalps, noses and ears.

My best advice for men is to remember to apply sunscreen when spending extended periods of time outdoors such as when biking, golfing and sailing, and to try our Visco-Elastic Transforming Serum to tighten the skin and even-out the skin tone. The first sign of aging in a man's face is in the lower eyelids and on the side, where they develop crow's feet. I've been using the serum for the past seven years and when I look back at photos from 10 years ago, my skin hasn't aged very much. And by using the serum on and around my eyelids, my skin elasticity has increased by about 50 percent.

Part III: Digging Deeper

"Let me tell you the secret that has led to my goal. My strength lies solely in my tenacity."

—Louis Pasteur

Chapter XI

Hemayet and Fenugreek

I first became aware of fenugreek in the 1970s when the natural food movement was getting started in North America, awakening many to the health risks associated with a steady diet of steak and potatoes. Back then, the seeds of this annual herb were popularly brewed as a caffeine-free tea. The roots of this plant go back much further into ancient times. The Greeks and Romans were known to have fed the plant to their cattle, a practice still prevalent in many parts of the world today. Remains of fenugreek seeds were discovered in Tutankhamen's tomb, and even earlier evidence has emerged in Iraq where some charred seeds were recovered and radiocarbon-dated to 4000 BC.

Fenugreek—also known as "Greek hay"—is native to the Mediterranean and parts of Asia. The plant produces small white flowers and grows to a height of about three feet with leaves shaped like elongated clover. Each of its sickle-shaped pods contains up to 20 or so amber-colored seeds, which have a slightly bittersweet taste that is sometimes compared to maple sugar (roasting reduces their bitterness). Fenugreek seeds are irregularly shaped and much harder than a sesame seed, making them somewhat difficult to grind. They are a popular ingredient in curry spice, so it comes as no surprise that India is today's biggest producer of fenugreek for human consumption.

I grew up eating a great deal of curry because my dad was from that exotic subcontinent, but I didn't specifically know about fenugreek back then. Nevertheless, it came as no surprise to me that the ancient Egyptians were the first to document some of the earliest-known medicinal applications of the plant. The more I researched fenugreek, however, the more I questioned whether it was the true ingredient in the recipe. According to the ancient formula's instructions, the hemayet fruit had a thick skin or shell that must be crushed. Hemayet must also be subjected to husking and threshing before proceeding with the recipe. These two processes didn't seem to apply to fenugreek or any other herb I could think of. Another disparity was related to what Breasted found when translating Case 45 of *The Edwin Smith Surgical Papyrus,* "Bulging Tumors on the Breast." The hieroglyph for "bulging" closely resembles the hieroglyph for "hemayet," so Breasted hypothesized that the fruit might have appeared as a bulging shape to the ancient Egyptians. No part of the fenugreek plant, from its small yellow seeds to its wispy green leaves, fit this description.

Something wasn't adding up.

I returned to Manniche's book where I first spotted the reference to *hemayet* to look for further clues. Above the section where Manniche wrote: "…*hemayt* is taken to mean fenugreek." Once again, it seemed the complexities of this ancient language were coming into play. But could hemayt refer to two different parts of the same herb? Or, were they two entirely different plants? Were they like grapes and raisins? The same fruit in two different states? Or, was Dioscorides, the ancient Greek pharmacologist, simply mistaken or misunderstood by later scholars?

The more I read, the more discouraged I was becoming. I wasn't ready to throw in the towel over fenugreek, but I had my doubts and the confusion only fueled my desire to solve the mystery of hemayet. I couldn't believe the answer hadn't been discovered after all these years. As far as I was concerned, the search was still very much on.

Not long after returning from our trip, we received information from the same travel agency that organized our first excursion about a new trip planned for 2006. Designed as a fundraiser for the Theban Mapping Project, the tour would be lead by Dr. Kent Weeks.

The 17-day trip would also include lectures and site visits led by Dr. Mark Lehner at the Village of the Workers on the Giza Plateau; a tour by Dr. W. Raymond Johnson at the Luxor Temple; an after-hours visit to the Egyptian Museum in Cairo; and a meeting with Dr. Zahi Hawass at his office of the Supreme Council of Antiquities. This trip provided a rare opportunity to meet several of the world's leading Egyptologists. Without hesitation, Joni and I cleared our calendars and booked reservations to join the tour. And once again, the boys were more than willing to miss a few weeks of school.

Chapter XII

Return to Egypt
January 2006

T he focus of our second trip to Egypt was on the Theban Mapping Project in the Valley of the Kings. Our first full day of the tour brought us to Saqqara, the city originally built as a necropolis for kings and nobles of the Old Kingdom (3100 to 2890 BC) close to where we saw the early Red and Bent pyramids the year before.

Built for Pharoah Djoser, who ruled Egypt from 2667 to 2648 BC, the pyramid was constructed in six graduated tables, or "mastabas," that resemble a massive staircase. The Step Pyramid is significant because it reflects advances in and techniques for building stone structures to greater heights than previously attained. But what I found most interesting was this pyramid was designed by Imhotep, whom James Henry Breasted speculated might have been the original author of the *Edwin Smith Papyrus*.

Sometimes referred to as history's "first genius," Imhotep was thought to have practiced as an architect, astronomer, poet and philosopher as well as a royal adviser to Pharaoh Djoser. It seemed he was a renaissance man before there was a Renaissance. So revered were Imhotep's contributions that he was one of only a few non-royals worshiped as a deity by the ancient Egyptians many centuries after his death. He was also considered to be history's first doctor. Most likely this prompted Breated's attribution, though Egyptologists today can find no evidence that Imhotep ever practiced as a physician. In any case, we passed outside a new museum dedicated to Imhotep as designer of the Step Pyramid. Built to showcase treasures discovered in Saqqara, the museum was scheduled to open later that year.

Dr. Kent Weeks is a tall, broad-shouldered and studious man who pauses to think carefully about what he wants to say. He was fairly quiet, but would generously share his knowledge at any time. Our primary contact with him was each evening when he and his wife gave an hour-long presentation that provided more detail about what we had just seen, and what we were about to see the next day.

Following the tour of Saqqara, we returned to the Great Pyramids on the Giza Plateau, where we wrapped up our time there with a rare tour inside the walls enclosing the Sphinx, an area not open to the public. Buried up to his shoulders at the time of Napolean's invasion in 1799, the Great Sphinx was not fully excavated until the early 1900s when the rest of its 200-foot-long body was revealed. (Later, I obtained several prints by David Roberts, a 19th-century Scottish artist that beautifully depict the partially buried Sphinx.) Submersion in sand most likely protected the ancient statue from serious erosion, and efforts have continued ever since to restore and protect it from further deterioration. But noticeably absent is the Sphinx's nose. The most persistent but inaccurate explanation for why it is missing places the blame on Napoleon's soldiers, who reportedly fired a cannon at it. Writings that pre-date the French invasion, however, describe a nose-less statue, so most Egyptologists agree that the *"rhinectomy"* likely occurred a thousand or more years earlier.

The following morning, we flew south to the coastal city of Hurghada to tour Fort Quseir, once a busy port on the Red Sea. We then proceeded to Wadi Hammamat, a dry riverbed that served as a major overland trade route between the two critical bodies of water connecting Quseir on the Red Sea and Thebes on the East Bank of the Nile. Stopping for brief hikes, we were shown some of the carvings on stone markers and cliffs recording thousands of years' worth of treks made by pre-dynastic hunters, ancient Egyptian sailors, Roman centurions, Christian monks and traveling Bedouins. Once again, as a North American traveler, my sense of time was deeply rocked as my feet kicked along the same sand and gravel these ancient travelers once trudged.

During one of these stops, Dr. Weeks directed our bus driver to make a sudden halt at what appeared to be a barren, featureless spot along the route. We were then escorted up a small path to a rock heap that appeared like any other, only to find it was covered with 12,000-year-old petroglyphs, rock carvings. Men running with spears poised for a strike were chasing stick figure drawings of gazelle-like beasts, which recorded the exploits of the pre-dynastic hunters.

Piling back on the bus, we stopped 10 minutes later to see two hills facing each other. Perched on the summits about 250 feet up were stone lookout posts, where the Romans once manipulated mirrors to send signals from post-to-post over hundreds of miles. In a matter of minutes, they could relay messages to warn of oncoming armies or monitor their own caravan position—an ancient form of "texting." Another short bus ride brought us to another fascinating area, again seemingly barren to the untrained eye. But as we studied the rock walls more carefully, we realized we were surrounded by hieroglyphs left by ancient Egyptians who had also traveled the Wadi Hammamat from the Red Sea to the Nile on their way to Lower Egypt or into the Mediterranean. Oddly, much of it was pornographic, showing pharaohs with erect penises.

When we finally reached Thebes, we spent the night at Luxor just across the Nile. Known as the "World's Greatest Open-air Museum," the region is home to a myriad of monuments, including Luxor Temple. Beautifully lit by evening, we toured this temple the next day with Dr. Raymond Johnson, field director of the Epigraphic Survey. Started in 1924 at the Oriental Institute of the University of Chicago, the Epigraphic Survey's mission is to produce photographs and drawings of the inscriptions on Luxor's monuments for publication. It has since been expanded to include conservation efforts. It was founded by none other than James Henry Breasted.

Luxor Temple is massive. Its complex of buildings stretches for what feels like at least a mile, and almost half as wide. Construction began under the reign of Pharaoh Amenhotep III around 1375 BC but continued over the millennia. As a result, Luxor Temple is the only religious monument in the world that includes Pharaonic, Macedonian, Ptolemaic, Roman, Christian and Islamic constructions. The night following our tour of the temple, I had trouble sleeping. But I knew it wasn't jet lag, sleeping in a hotel bed or the delicious

Egyptian cuisine I was probably over-indulging in. My restlessness was definitely in anticipation of the next day's adventure. At last we would return to the Valley of the Kings.

The Symbolism of Pyramids

The shape of pyramids is thought to represent the primordial mound from which the Egyptians believed the earth was created, according to an online source. The shape of the pyramid is thought to be representative of the descending rays of the sun. Most pyramids were faced with polished, highly reflective white limestone to give them a brilliant appearance from a distance.

While it's generally agreed that pyramids were burial monuments, there is continued disagreement on the particular theological principles that might have given rise to them. One suggestion is that they were designed as a type of "resurrection machine."

The Egyptians believed the dark area of the night sky around which the stars appear to revolve was the physical gateway into the heavens. One of the narrow shafts that extends from the main burial chamber through the entire body of the Great Pyramids points directly towards the center of this part of the sky. This suggests the pyramid may have been designed to serve as a means to magically launch the deceased pharaoh's soul directly into the abode of the gods.

Chapter XIII

Valley of the Kings

The Valley of the Kings is a valley in Egypt where for a period of 500 years (from the 16th to the 11th century BC), tombs were constructed for the pharaohs and powerful nobles of the New Kingdom. The valley is located on the west bank of the Nile. The area is also known as "Kings Valley" or KV, which is part of a numbering system devised by the Egyptian Antiquities Service to document the region's tombs.

The tomb of Pharaoh Seti I is KV17. Once again with special permission, we were able to tour KV17, the tomb of Pharaoh Seti I and view the vividly painted wall art. The build-up of humidity and perspiration generated by thousands upon thousands of tourists is gradually eroding the soft limestone walls of this and other tombs in the Valley of the Kings. Plans are in place to install cool lighting systems, provide ventilation and stagger visits into the evening to mitigate the accumulations of this human-generated moisture. In light of the risks to these irreplaceable antiquities, I was grateful and excited to have the opportunity to return to Seti's tomb. Something about the tomb stirs me beyond my five senses and into the realm of the sixth. The experience is entrancing and time altering in a way that is difficult to describe.

During our visit to the tomb, I was curious to see if the hat I'd lost deep down in the mini-moat the year before was still there. Much to the great amusement of my two young sons, there it was.

Further along in the burial chamber, the one with the lapis-blue night sky painted on the ceiling, I was once again mesmerized by this astonishingly beautiful artwork. Flash photographs are another potential source of damage to the paint, so I enjoyed every moment of this visit, committing the images to memory. They didn't want anyone to take any kind of photographs, which is why we have no photos of the inside of any tombs except for Nefer, who was considered the Elvis of his time, and who sang for pharaoh Djoser.

Next, we toured KV5, which was reportedly passed over in the early 1800s when an English explorer digging near the entrance abandoned his efforts, thinking nothing more was there. But it was an understandable mistake. Every 100 years or so massive flash floods cause torrents of water to wash down from the mountains above the Valley of Kings, depositing silt into any and all openings along the way. As a result, the entryway to KV5 became filled from bottom to top with impacted sediment, making it appear to be another wall of stone. But in 1995, Dr. Weeks and his team took a second look, based on an earlier English explorer's writings. Brushing aside a handful of debris at a time from the ceiling of a hallway, they created an opening just large enough to crawl inside and made what is considered the biggest discovery in the Valley of the Kings since the celebrated find of Tutankhamen's tomb in 1922.

Because it's still under excavation 15 years after being discovered, KV5 is not yet open to the public. As I was working on this book, I learned from Dr. Weeks that the process of preservation and its sheer size is the reason it's taking so long. With more than 100 chambers for the 52 sons of Pharaoh Ramses II, KV5 is the largest tomb ever discovered in the Valley of the Kings.

Our group was treated to a private tour with Dr. Weeks, who gave us an amazing look inside this extensive underground burial site. With the exception of a carving of the god Osiris at the end of the long central hallway, KV5 does not have much in the way of wall or ceiling paintings. Apparently, you had to be a king or queen during the time to warrant such elaborate preparation for the afterlife. It was anticlimactic to not see beautiful art on the walls after seeing the other tombs. But to discover a tomb thousands of years in an age so extensively studied was a feat of itself.

I was hoping to ask Dr. Weeks about hemayet during the tour, but never found the right moment. He was busy focusing on his work, conducting the tour, and answering questions from our group about various details and artifacts. I didn't want to derail the flow. I made a mental note to ask him later. I was becoming skeptical that I'd ever find the answer.

As we continued the tour, we learned that one ruler who reached the pinnacle of reverence to receive wall paintings was Queen Hatshepsut. We visited a dramatic monument constructed in her honor just east of the Valley of the Kings. Considered one of the major showpieces of Thebes, the Temple of Hatshepsut is stunningly situated below one of the area's tallest peaks. The rear portions of the building were carved into the rock.

Construction on the mortuary temple began during the queen's reign from 1479 to 1458 BC, just after the *Edwin Smith Papyrus* was thought to have been written in 1500 BC. I wondered about this timing from the standpoint of the bitter melon fruit I had researched before our arrival. One of Hatshepsut's major accomplishments was to establish trade with other lands, specifically that of Punt. No longer in existence as it was in ancient times, Punt is generally thought by scholars to have been south of Egypt. Hatsheput ruled later than the Edwin Smith Papyrus was written, but did trade with Punt also happen before her reign? If so, it was one way the bitter melon might have come to ancient Egypt.

That evening, we visited a souk in Luxor. This time, I had my sights set on the produce vendors for anything that resembled the bitter melon fruit. Melons were indeed everywhere, but none that looked anything like the distinctive bitter melon, bulging or flat. I was beginning to have doubts about this particular fruit. Would a large melon like that be "threshed like grain" as the "Recipe for Transforming an Old Man into a Youth" instructed? It seemed highly unlikely.

We returned to Cairo the next day, and Dr. Hawass arranged a special tour of the Egyptian Museum for our group, which opened at its current location in 1902. Designed by French architect Marcel Dourgnon in the neoclassical-style, the peach-colored museum on Cairo's Tahrir Square contains more than 100 halls. The first floor is devoted primarily to the largest statues. Upstairs we found the smaller artifacts, including jewelry, coins, pottery, cosmetics containers and other everyday items used by ancient Egyptians. The second floor included the treasures of Tutankhamen and the mummy rooms.

I had been looking forward to this with great anticipation. In this special hall are the remains of 11 pharaohs and queens. A second hall has since been opened to house additional royals, including Queen Hatshepsut, whose mummy was identified in 2007 by Dr. Hawass and his team through DNA testing of an abscessed tooth buried separately in a canopic jar. (A small jar, containing the organs of the pharaoh.) We observed the great rulers in hushed reverence. I saw Ramses II, Ramses III, Thutmose III, Amenhotep III and his wife, Queen Tiye as well as my favorite pharaoh, Seti I, whose tomb in the Valley of the Kings now appeared to permanently include my hat.

Seti I was the son of Ramses I and Queen Sitre, and father of Ramses the Great (II). Unlike other royals who lived to be middle-aged and sometimes older, Seti I appears to have lived no more than 40 years, the average lifespan of an Egyptian at that time. His five-foot-seven-inch mummy shows no signs of trauma, nor do any records indicate a cause of death. He reigned sometime between 1290 and 1279 BC, about 200 years after the *Edwin Smith Papyrus* was written. I mention this only because Seti looks as if he might have taken advantage of the "Recipe for Transforming an Old Man into a Youth," at least his spectacular remains do. Widely regarded as the best-preserved mummy of all of ancient Egypt, Seti's mummy, with its impossibly unwrinkled face, almost looks as if he might suddenly come back to life at any moment. I've passed older looking men on the sidewalk.

As in many of the tombs, photographs aren't allowed in parts of the museum, so I came prepared with a small pad of paper from our hotel room. I wanted to sketch some of the pharaohs to create a comparative "royal" facial analysis. Looking at them from a surgical perspective, I captured their features relative to one another as well as the profile of their noses. Aside from Seti, I found the face of Ramses II to be the most interesting because an X-ray of his skull had revealed little seeds packed inside his nasal cavity to maintain its dorsal height. This would prevent the nasal cartilages from collapsing and altering his profile over the centuries to come, a sort of post-mortem form of facial plastic surgery to keep the great pharaoh looking as much like himself as possible in the after-life.

King Tutankhamen's mummy is not in the museum. It is displayed in its original tomb in the Valley of the Kings. I would've loved to make a plaster mask of his features and those of the other pharaohs, but knew the process of smearing Vaseline on the mummy's face would not be appreciated. I would have to prove that the next stages of the process would not damage it in any way. I wondered if I could perhaps mask one of the skulls, but it would be missing the nasal cartilage and other tissues preserved through mummification, the very tissues that distinguish a face from a skull. Could I perhaps first practice on the face of a lesser-known mummy, or one that was declared too decayed to save? I wondered.

That night as I was drifting asleep, my mind raced, thinking of all the ways I might accomplish my mummy-masking dream, but I knew that even if I was granted permission, there wouldn't be time. We would soon be leaving Egypt, but one last adventure awaited us. This next adventure would take us to a monastery that supposedly had a well-stocked library. I hoped I might find information about the hemayet fruit, but to get there, we would have to travel the time-honored, four-legged way.

The Art of Masking

When a facial plastic surgeon performs a rhinoplasty—more commonly known as a "nose job"—the traditional method of documenting the change is with a "cross-table lateral," a photograph that captures the patient's profile. I've probably performed more than 2,500 rhinoplasties during my career.

During one surgery, I came to the realization that a photo can only provide a two-dimensional record of the rhinoplasty. Wouldn't a three-dimensional version be much better? I considered making a mold of the nose, but decided a mask of the entire face would provide a more comprehensive view. Because I had no idea how to mask a face, I arranged a visit to the College of Arts and Crafts in Oakland, California, where an elderly gentleman instructed me in this unique art.

First, a person's eyelashes, eyebrows and hairline are coated with Vaseline to allow for easy release of the silicone mask. Next, straws are inserted in each nostril, preferably the fast-food type intended for milkshakes, which are a little wider than the standard variety. This allows the person to take in air more easily yet doesn't distort the nostril shape in most individuals. And finally, a two-part silicone gel polymer is mixed together and carefully poured over the person's entire face to form a "polymer negative" mold. In about 15 minutes, the polymer fully hardens. A plaster material is then laid in overlapping strips like a mummy across the more flexible polymer mask to bolster its shape. The plaster part, or "mother mold," is removed about 10 minutes later once it hardens. The polymer negative is also removed and slipped inside the mother mold for support. I have created masks before, during and after facial surgery to give patients a glimpse into facial transformation. I hold the U.S. patent on facial masking in surgery. U.S. patent dated November 11, 2003 and titled: "Method for making face masks for reconstructive surgical patients."

For the person being masked, the process is over. The straws are removed, and the face degreased. The process is simple and easy, unless you're claustrophobic, fidgety or have a round of golf to play.

The final stage is carried out later. Hydrocal, a special white gypsum powder, is slowly mixed with water until it forms a thick sludge. Once the straw openings in the nostrils are plugged with putty, Hydrocal is poured into the negative mold as it is still cradled in the mother mold. In a few hours, a likeness of the person is pulled from the negative, showing every wrinkle, scar and the individual's precise features. The white plaster mask is so life-like that one patient called it "spooky." Nevertheless, the technique provides an exact replica to work with, which I create before, during, and after masks for certain selected surgical patients that will show marked three-dimensional changes.

Chapter XIV

Mount Sinai

A random chorus of strange groans greeted us at about two o'clock in the morning when my family and I approached a holding area for camels at the base of Mount Sinai. In the dead of night, the noises made by camels, sound much eerier than by day. Whatever it is these curious creatures of the desert were attempting to communicate, whether to each other or to us, the sounds seemed more what I imagined a drowning horse would make.

We had arrived at the holding area in the middle of the night in preparation for our journey up the mountain to watch the sunrise. On the way back down, we planned to stop at Saint Catherine's Monastery to visit its library of ancient books and scrolls.

Mount Sinai, located near the end of Egypt's Sinai Peninsula about 200 miles southeast of Cairo, is the summit where Moses legendarily received the Ten Commandments. References to the rugged peak can be found in the Bible and throughout the Quran as well. It isn't the tallest of the region's red granite mountains, but certainly it's the most famous. At an elevation of almost 7,500 feet, Mount Sinai is barren, rocky and windy. And at night, it's unusually cold. So cold, in fact, that I went to bed fully clothed the night before in the old stone El Wady motel where we stayed before our ascent.

I glanced at Joni and the boys. We had bundled ourselves

almost past recognition for the trip. The boys were layered in sweatshirts, down jackets, hiking pants and warm socks. And though they were both a bit sleepy, they slowly rallied once they saw the camels we would be riding. In those early morning hours, it was still icy cold and so dark that our only points of reference were the flashlights we each carried to illuminate our way as we followed our Bedouin guides to four single-humped dromedary camels tied in sequence by ropes about eight feet long.

Unlike horses, camels are mounted when they are seated. They rise first from their back legs, unfolding their flexible front legs at the knee to fully stand. Joni rode the lead camel, followed by Austin, me, and then Spencer. The two guides walked, one at the front, the other from behind following Spencer. We were aware of other small groups like ours, climbing up the mountain with us that morning, but only because we heard their voices or spotted the flickering of their flashlights as they moved.

We had previously ridden camels in Saqqara, so the sensation of being in the saddle was a familiar one. What we weren't quite ready for, was the feeling of sensory deprivation. The soft feet of our camels made almost no sound as they touched the ground, so unless one of the leggy beasts let out a moan or one of us spoke, our ascent up Mount Sinai was not only made in complete darkness, but mostly in complete silence as well. There was no reassuring glow of streetlights, store lighting, porch lights or the like.
Only rarely do we find ourselves in parts of the world where light is all but nonexistent at night, not even spilling into the sky as glare from neighboring cities.

"Dad?" It was Spencer, his small voice, sounding worried, wanting to know I was still there in the dark of the night.

"Hey Spence, how are you holding up back there?" I asked, wanting to reassure him.

"Good."

"Did you know that to measure light pollution, scientists use the Bortle Dark-Sky Scale?" I asked. Spencer was not quite nine years old at the time, and I knew he was a little freaked out by the experience.

"No."

"Do you want to hear more?"

"I guess."

I'm not so sure he was intrigued with the science of the Bortle Dark-Sky Scale, but I kept the flow of dialogue going in an effort to let him know he was safe. "It ranges from Class 9 in a metropolis like Tokyo, Japan, where almost no stars can be seen, to Class 1 in remote places," I continued. "Do you know what I mean by remote places?"

"Not really."

"Like the Australian Outback, the Siberian wilderness or out at sea at least 300 miles from land in any direction, where no artificial light can obstruct the sky."

"Okay."

Spencer seemed to be more at ease, so I took a break from the science lesson and looked around. Without a doubt, we were as close to being under a Class 1 sky that morning as I had ever experienced. Only when I shined my flashlight ahead to see Austin or behind to see Spencer could I see much of anything. When I pointed the beam to my right, I could see the mountain or a rocky outcropping. When I aimed the beam to my left, however, I saw nothing, as if a black hole had swallowed the beam. I wished I could see where we were going, and looked forward to seeing the path by day when we made the trip down later. (Later on our descent, I learned, oh, be ever so careful of what you wish for.)

Along the way, we saw two nuns, making a religious pilgrimage together. They looked to be in their 60s and were wearing long black dresses and scarves. I was immediately concerned for them because they were traveling alone by foot without a flashlight or guide to help them. I called out—afraid they might fall off the mountain. We learned they were from Romania, but there was a language barrier, and it was impossible to communicate. Concerned they might fall, I gave them a flashlight and we continued on our way.

Looking up, the view was spectacular. The Milky Way, shooting stars, and constellations were all visible in such clarity that I felt as if I were seeing them for the first time. No wonder ancient Egyptians communed so well with the night sky, becoming expert astronomers. I couldn't help but think of my favorite starry, blue painting on the ceiling of Seti's tomb, which suddenly came alive for me in a whole new way.

Two hours later, we were nearly at the summit, when we made the customary tourist stop for tea and packaged snacks at a little stone shop operated by Bedouin merchants. We had to squint against the warm light inside, but it was a welcome trade-off after the cold darkness.

Refreshed and back in our saddles a short while later, we arrived at the peak of the mountain around five o'clock in the morning. As the sky began to lighten in its seamlessly imperceptible way in pearly shades of pink and blue, we realized we were in a group of about 30 people, who had come by camel as we had. There were other visitors who had hiked. There was a low hum of conversations in subdued tones, perhaps from morning grogginess, but also reverence. The moment of spiritual awe came at the very instant the sun crested over the horizon.

"Beautiful, isn't it?" Joni, whispered.

The four of us sat together silently, watching from the peak of the mountain. The moment was not lost on any of us. Together, yet alone with our thoughts, it was not a stretch to believe something miraculous had indeed happened long ago on this sacred peak.

Later, with stunning views of Israel, Jordan and Egypt as our backdrop, we began our descent to Saint Catherine's Monastery near the base of Mount Sinai. Now I realized why the beam from my flashlight had dissipated into infinity when I shined it to one side on our way up. Cliffs, precipices, steep drops. *Yikes!* They all loomed off the trail so severely at points that it forced us to look away and tighten our grips on our camels. For some strange reason, the animals preferred to walk on the outside of the trail closest to the edge. Perhaps Spencer had a better sense of our precarious situation on the ride up, while the rest of our family had been blissfully unaware of any danger. *Why would the camels walk so close to the edge?* At one point, I summoned the courage to peek over a cliff. Not seeing any camel skeletons below, I reassured myself that these creatures knew what they were doing.

"Mum, is this the way we came up?!" Austin asked.

"Is this safe?" Spencer asked, peering over the edge. "It doesn't look safe."

Joni turned around and shot a look at me. I could tell by the expression on her face that her mother bear instincts were now on full alert. "*Ay yi yi!*" She mouthed.

Ay yi, yi, indeed. We hadn't been fully aware of what this leg of the journey entailed when we signed up.

"These camels have taken this trip time and time again. Have faith," I said to my terrified family, in a calm, authoritative tone, trying to reassure them as well as myself. "These are sure-footed, sturdy creatures whose ancestors upon ancestors knew this terrain better than any human."

Austin turned around on his camel and shot me a skeptical look.

"*Gggghaughhhh!*" A camel nearby suddenly bellowed this new, even more bizarre noise, a loud gurgling with a wail. I looked just in time to see a large, pinkish-red fleshy pouch hanging out of the side of its mouth, which it retracted after a few seconds. *Was it sick? About to keel over?* Our guides didn't seem at all concerned, and the camel continued downward as if nothing had happened. We later learned that male camels possess a "dulla," a type of bladder they can toss out of their mouths to display dominance. I realized making this trip in utter darkness had been far more peaceful.

Saint Catherine's Monastery was a welcome sight. We arrived there mid-day. On the last leg of our challenging descent, we were treated to a bird's eye view of the complex built from the abundant red granite. Within the monastery's fortress-like walls, lay a cluster of buildings constructed over the centuries in an amazing hodge-podge of Byzantine, Gothic and other early styles of architecture. Built under the order of Justinian I, the Byzantine emperor who reigned from 527 to 565 AD, the monastery was designed to enclose the smaller Chapel of the Burning Bush constructed about 200 years earlier. Once inside, we indeed saw a living bush, a type of bramble with small green leaves that is believed to be a direct descendent of the one through which God communicated to Moses, instructing him to lead the enslaved Israelites out of Egypt.

The 25 monks who live on the grounds of Saint Catherine's, the world's oldest continuously inhabited monastery, are Orthodox Christian monastics, whose common language is Greek. The local Bedouin are Arabic-speaking Muslims whose ancestors traversed the

Sinai Peninsula. Together, the two diverse communities have lived in harmony since the sixth century, and still meet weekly to bake bread at the monastery. Two structures can be seen rising above the walls of Saint Catherine's side-by-side: the dome of a small mosque and the bell tower of the Basilica of the Transformation, the monastery's main church.

The Archbishop Damianos of Sinai, who presides over Saint Catherine's, once wrote this passage: "Where difference of language, religion, and culture might have caused tensions and discord, there has been, instead, a relationship of deep regard and mutual support between the monks and the Bedouin; each group respects the other, in religious matters above all. This, indeed, sets an important example for those who seek peace in our times."

Much to the amusement of my family, the only other inhabitants of the monastery appeared to be a colony of cats, less concerned with world peace, of course, and more concerned about warming themselves in the sun. During our travels, I had patronized what seemed like every bookstore we passed, and each stop was met with groans from Austin and Spencer. Their usual chorus was, "Not, again!" And was followed by the usual eye-rolling and complaints, typical for boys their age. It wasn't so much that they found the books boring, just heavy. Though they may have felt like pack mules, we all took turns carrying shopping bags of books and I would remind them how the ancient Egyptians moved giant blocks to build pyramids in the hot sun with no air conditioning to come home to at the end of the day. In any case, today there were no bookstores, just an ancient library with ancient scrolls and books that were not for sale and cats to amuse them along the way.

We were escorted to the library by one of the monks through a maze of corridors and stairs to a small, dimly lit room. A bearded monk, wearing a flowing black robe and mitre hat rose from his chair.

I extended my hand. "*As-salamu alaykum,*" I said in Arabic, which is the customary greeting and means "peace be upon you."

"Do you want to see the library?" He asked in perfect English with a Texan drawl.

I was startled and couldn't help but smile. I certainly hadn't expected to encounter a fellow American here, especially one with a southern drawl. "Yes. This is my wife, Joni, and my sons, Austin and Spencer."

"I'm Father Justin," he said.

"You're from the States?" I asked.

"Yes," he said, and we learned that he was the only American-born monk at Saint Catherine's.

As Father Justin opened the padlocks and parted the ancient wooden doors, we were greeted by a rush of air laced with that familiar scent of old books—extremely old, in this case. We stepped inside to see a treasure trove of ancient scrolls, silver cased texts in Persian, maps of the known-world from hundreds of years ago and early printed books. Indiana Jones would have been impressed.

"How many books are in here?" Austin asked.

"About 5,000 in all," Father Justin replied.

I looked around. The walls were lined in shelves two stories high with scrolls, most of which looked as though they had never been unrolled. I'd been hoping to find something medical in nature about the hemayet fruit, but I quickly realized that would not be possible. Even if a scroll or other document that unlocked the Egyptian pharmaceutical mystery was here, it would likely be in an ancient language, as Father Justin explained, taking one or more scholars to translate, transliterate or both. Still, I couldn't help but wonder if the hemayet answer was within the walls of this very room, tucked inside a single scroll. I might be standing just inches away from the one word that, if properly translated, solved the mystery. If only the secret could have been unlocked as easily as the padlocks to the library itself.

I knew this would be a mammoth effort and pointless. Father Justin was kind with his time and shared the literary masterpieces in the monastery's collection, rivaled only by the Vatican in size and scope.

One of St. Catherine's prize pieces is a fragment of the Codex Sinaiticus, kept in a special glass case. Dating from the fourth century, the Codex is considered one of the earliest copies of the Bible in existence. The monastery is also known for its extensive collection of religious artifacts and some 2,000 early icons, including a colorful sixth-century depiction of Christ painted in pigmented beeswax on wood. At the time of our 2006 visit, Father Justin was organizing a collection of the monastery's icons for an upcoming display at the Getty Museum in Southern California. A generous man, he took us to see the gilded panels he had assembled, which were graced with saints, angels and other religious figures.

"Thank you," I said, extending my hand again to our Texan-speaking monk. Today, was strike three in my search to decode the mystery of hemayet, but experiencing this ancient library was an event we would remember for years to come.

Outside, the air was still crisp and cool, but we followed the lead of the resident cats and stayed in the sun as much as possible. The camels in the holding pen were now visible to us in the daylight, about 25 of them, their wails sounding much less unsettling now that we could see their faces. Depending on the angle and the darkness of the markings around their lips, it looked to me as if some even appeared to be smiling. My guess is they were probably happy to have us off their backs.

When we returned to the old stone motel, our driver, Moisen, was ready to take us through the Sinai Desert and across the Suez Canal to Cairo. Moisen was strong, muscular and his neck nearly equaled the width of his head. I never tired of watching him toss our heavy suitcases onto the roof of the Jeep with one hand. Accompanying us on this particular part of our tour was a younger man in his twenties who wore a dark gray suit with an Uzi sub-machine gun tucked inside his jacket.

Their presence was a bit disquieting and intimidating at first, but we'd all become familiar with the routine in Egypt. Tourism is an important industry for Egypt, so the government wants to make sure visitors are always safe. Once, during our first trip to Egypt, we made a stop at the famous Hanging Church in Cairo, arriving to see dozens of police officers lining the streets in front of steel barricades. "Why all the security?" I asked our guide that day, thinking their presence was for a visiting dignitary.

"It's for you," he had said, meaning all visitors.

Terrorist attacks against visitors are rare, but there had been a massive setback in tourism following two separate incidents in 1999. Sadly, one ambush took place inside Queen Hatshepsut's beautiful tower in Luxor. Since then, the government stepped up security for tourists, not wanting a repeat of any kind. Now, since the Arab Spring, tourism is sadly at an all-time low. On the other hand, for those willing to visit Egypt, the lines are short and you'd have the tombs almost to yourselves.

As we climbed into the Jeep that afternoon, I thought back to our gracious Bedouin guides, Father Justin and the other Greek Orthodox monks we had met earlier that day. Yes, peace in our times as experienced by Saint Catherine's would be a great gift to all. In the meantime, we knew we were safe with Moisen and our young protector. Rumbling along in the Jeep on our way back to the Four Seasons Hotel in Cairo, Moisen kept up a brisk pace, wanting to be home in time to see Egypt's soccer team compete in the annual African tournament. I sat next to Joni, looking out the window. We passed a few oil derricks in the desert, silent signals that we were gradually entering Egypt's modern-day world.

The next morning, we arrived at Cairo International Airport to begin the trip home, far away from the monuments and antiquities that had transported us back in time over the past few weeks. As the jet lifted off the ground, I felt the same sense of sadness I experienced the first time we left Egypt the year before. After two trips, Egypt was beginning to feel like home in a most ancient and ancestral way. And though I had enjoyed every moment of our two trips to Egypt, I couldn't help but take stock of my efforts to figure out what hemayet fruit was in the "Recipe for Transforming and Old Man into a Youth." It had been six years since I had stumbled upon the recipe, and I still had both feet firmly planted on home plate, swinging at air and missing each ball thrown my way.

Chapter xv

Pearl of the Mediterranean
Alexandria, Egypt – Summer 2007

A lthough I was no closer to deciphering the meaning of hemayet, a chance encounter seemed to support my efforts in developing a skin cream once again. I was in Southern California presenting wound healing research on a type of energized water, when by chance I met Tom Hrubec, director of product development of Grant Industries, a family-owned company that provides specialty materials for the cosmetic industry. Located in New Jersey, Grant Industries counts among its client companies such as Revlon, Avon and Estee Lauder.

Tom was impressed with the research I'd presented, but during a break he asked, "Have you ever considered developing your own skin care line?"

I smiled, of course. "As a matter of fact, yes."

"I'd like to introduce you to a friend of mine," Tom said. "His name is Jules Zecchino. He worked at Estee Lauder for 20 years and was once the chief chemist of research and development." I learned that Jules had developed Ceramide Capsules for Elizabeth Arden as well as Vaseline Intensive Care early in his career. We exchanged information and agreed to meet later.

Meanwhile, the plan was to continue my search for the meaning of hemayet at home in California. We had met so many Egyptologists and other knowledgeable people during our trips that I began corresponding with them by phone and email, hoping to find an answer. But as luck would have it, another opportunity to travel to Egypt unexpectedly presented itself.

I was invited to give a lecture at the American Academy of Facial Plastic and Reconstructive Surgery (AAFPRS) with Dr. Bill Silver, a facial plastic surgeon from Atlanta. I was to present my technique to create plaster masks of patients undergoing facial plastic surgery in Toronto, Canada, in the autumn of 2006. I managed to convince a colleague of mine, Dr. Michael MacDonald from San Francisco, to let me mask his face live for the audience. The day of the lecture, I met another surgeon, and another invitation was presented. Dr. Cemal Cingi, the event coordinator for the European Society of Facial Plastic Surgery, approached me after the lecture and invited me to present it again at their 2007 conference. The event would be held in Bodrum, Turkey, a veritable hop, skip and jump from Egypt. After traveling all the way from California to Turkey, it would have been crazy not to make the trip Egypt again. This time, however, we would travel in summer, a character-building experience for all of us.

Through our previous trips, we'd gotten to know Hatem Ali and his wife, Noha, who live in Cairo. Hatem worked with the Quest Travel Group, and he and his wife had two sons, Karim and Omar. During our previous trips, Austin and Spencer played with their children and we'd all kept in touch through Facebook and emails. We contacted Hatem and Noha and planned a vacation with our families to Alexandria. Desiring a less structured itinerary this time, we met Hatem and Noha and traveled with them to Alexandria where many vacationers from Cairo traditionally go during the summer to escape the heat.

We arrived in Alexandria on a beautiful, hot sunny day in June. Alexandria is Egypt's second largest city. (Cario is the capital and largest city of Egypt.) Known as the "Pearl of the Mediterranean," Alexandria was founded in 331 BC by the legendary conqueror Alexander the Great. The city later became home to one of the Seven Ancient Wonders of the World, the Lighthouse of Alexandria. Constructed around 260 BC, the top of the 400-foot stone tower was equipped with a reflective metal mirror by day and lit by fire by night to guide sailors into its harbor. For several centuries, it remained one of the tallest man-made structures in the world until a series of earthquakes finally toppled what was left of it into the sea in 1323 AD. It was that same succession of temblors and corresponding tsunamis that eventually knocked Cleopatra's palace into the Mediterranean as well. Alexandria is perhaps best remembered as "her" city.

In the summer, the corniche el Nil rimming the Mediterranean becomes so overcrowded with people crossing the beach and back that it's almost impossible to drive a car along the boulevard. The water was a lovely jeweled-green, though we did see litter floating. The water was warm, but still refreshing to be in, especially when the temperature is a blazing ninety degrees. Sea breezes help as well. On the beach, women wear colorful galabiyas in orange, turquoise and blues, a full-body Arabic dress they lift only to their knees when they're enjoying the water. The men are all in swim trunks and a fair number of women wear Western-style swimsuits as well.

"Do you feel out of place in a two-piece?" I asked Joni, as we were walking along the water's edge.

She laughed and shook her head. "No, I really don't. It's not as if I'm the only one here in a bikini."

"Good," I said. The locals seemed to gracefully accept this mixture of attire on their shores, and I was glad she didn't feel conspicuous.

We spent most of the afternoon on the beach with Austin and Spencer and stopped for ice cream. Later, we watched as the sun set over the Mediterranean Sea, and it was quite beautiful in dusky shades of orange and pink, but my thoughts were on the library of Alexandria and the French invasion of Egypt under Napoleon. I wondered what our knowledge would be if that library had not burned down. Would man have advanced at a faster rate? Or, would the Catholic church have intervened in scientific progress all the same? We dried off and changed and headed for the one place in town I thought might solve the hemayet mystery.

A gleaming marvel of a building, the Bibliotheca Alexandria, which opened in 2002, sits on the banks of the Mediterranean and features a circular, glass roof that slants toward the nearby Mediterranean shore. It is believed to be the same location where the ancient Library of Alexandria once stood. (See "The Library of Alexandria" on page 111) The building is covered with "eyelids," a series of six-by-two-foot triangular panels that automatically open and close to mitigate the sun. Inside, the building features 11 levels with numerous libraries, study rooms, museums, art galleries, a conference center and a dramatic planetarium near the entrance. With a startling collection of about 250,000 books, the Bibliotheca Alexandria plans to collect 5 million volumes of books, concentrating primarily on topics relevant to Egypt, the Middle East and Islam.

The new Bibliotheca Alexandria is a wonder to behold, fully computerized, networked and linked in every conceivable modern way. The librarian showed us a small portion of an ancient scroll, which she believed was from the original library of ancient times. I had this sense of excitement once again, as I searched through the ancient scrolls, thinking I'd finally find the meaning of hemayet, but I couldn't find any documents related to ancient Egyptian medicine. Once again, I was facing the daunting task to cull through a vast quantity of information that I faced at Saint Catherine's Monastery. No one at the library was familiar with the word and there was no way to expediently plumb the vast array of books, manuscripts, journals and scrolls sometimes written in ancient languages to find the word "hemayet" and its translation. I felt so close to the answer, standing in the library, and yet the meaning of hemayet remained a mystery. The irony of the situation was not lost on me.

The Library of Alexandria

Scholars have mourned the loss of the legendary ancient Library of Alexandria for centuries. The library was built in about 300 B.C., and then burned down around 80 B.C., and then again around 270 A.D. Historians believe the library once archived about 500,000 scrolls, documenting history and knowledge from the ancient world. The *Edwin Smith Surgical Papyrus* could have been in the library at one time. And if so, fortunately it was moved elsewhere.

Beginning in about 300 BC, ships entering Alexandria's harbor were required to surrender for copying any books they carried on board as a fee to harbor there. Knowledge was the trade. It's believed the library kept the originals, returning copies to the ships. As a result of the Alexandrian quest for knowledge, which included chartered trips to Athens and other cities to procure copies of their books, the library was well stocked. The Library of Alexandria housed the masterpieces of classical civilization, including the works of Aristotle and Plato; original manuscripts of Sophocles, Aeschylus and Euripides; Egyptian treatises on astronomy and medicine; Buddhist texts; and the first translations of the Hebrew scriptures.

I had little doubt the ancient Library of Alexandria would have held the secret to the hemayet fruit, if not an actual copy of the *Edwin Smith Papyrus.* And if the answer were not readily available, a fleet of resident scholars would have been available to help. But history's illustrious archive had vanished by about the 4th century AD, the record of its founding being clearer than any documentation of its demise. One account has Julius Caesar accidentally destroying a portion of the library during the Alexandrian War in 48 BC. Later accounts involve possible decrees to burn its books during a series of sieges and invasions, variously placing the blame on the Romans, Christians and Arabs.

Part IV: The Breakthrough

"Egypt became the cradle of scientific development. The progress it attained in astronomy, architecture and pyramid building is often emphasized, but undoubtedly its knowledge of medicine and surgery has been underestimated."

—Nabil I. Ebeid

Chapter XVI

Mystery Solved
Santa Rosa, California – 2008

I followed up with a few helpful leads after our trip to Alexandria, but all efforts led to dead ends. Meanwhile, I kept in touch with Tom Hrubec and Jules Zucchino through a series of phone conversations, and though I was no closer to unlocking the mystery of hemayet, we arranged a meeting at my office in Healdsburg, Calif., on February 11, 2008. I also invited a fellow physician, Dr. Ron Botelho, who is a close friend and good businessman. The goal was to get to know one another, discuss our combined knowledge of skin care and to consider forming a unique, new skin care company.

Jules grew up in New York and seemed every inch the New Yorker to me with his accent and especially his bold directness, which I appreciated. Always interested in how things work, I learned that Jules took his first chemistry class in high school and from that day forward he was hooked. He attended the University of Bridgeport in Bridgeport, Conn., as a pre-med student, and graduated cum laude. Later, his life took a detour when he interviewed for a position as a formulating chemist with Chesebrough-Ponds, and Jules knew immediately this was the career he wanted.

We were sitting around a conference table at my office, and I brought out my copy of the *Edwin Smith Surgical Papyrus* and showed them the recipe for improving skin. The men read the recipe with great interest, passing the book around the table and their reaction was much the same as mine had been the first time I'd stumbled upon the recipe. "Fascinating," one said.

"Could this recipe be replicated?" Another asked.

"Definitely worth exploring," one commented.

Then Jules looked up from the book and said, "I have always believed that the ancient civilizations had a lot more knowledge than we give them credit for."

And then, of course, they all turned to me with the next logical question.

"What's hemayet?" They asked, looking at me with curiosity and high expectation.

I sat back in my chair and had to laugh. "Well, that's the one problem. I don't know. I've been searching for the translation for the past eight years."

"Eight years?" Someone asked.

"Yes," I said, and I could tell they were taken aback.

"You mean you don't even know what the main ingredient is?" Jules blurted out. "It's been eight years. What makes you think you can find the ingredient?"

There was a long moment of silence, all eyes fixed on me. I smiled. "Don't worry. You don't know me very well yet, but I'll find out," I said, glancing around the table. Suddenly the excitement that had been building vanished and had settled into skepticism. Despite eight years of failure, I was committed to finding the answer and what they didn't know about me is that I never give up.

I continued my research for the next three months, then in May, I came across some materials I'd kept from our trip to Egypt two years earlier, including the phone number of Dr. Kent Weeks.

It was 11 o'clock in the morning, and it was three hours later in Connecticut where Dr. Weeks maintains a summer residence. I punched out the number and waited. After a few short rings, he picked up. "Hello?"

"This is Stan Jacobs," I said. "We met two years earlier on a tour of Valley of the Kings."

"Oh yes, I remember you and your family," he said.

We made small talk, but then I explained my reason for calling. "Hemayet?"

"Yes," I said, and spelled it for him, lest there be any confusion. "The recipe describes it as a fruit that must be 'threshed' and 'winnowed.'"

There was a moment of silence and then he said, "I don't know, but let me give it some thought and get back to you."

The next morning after performing a facelift in my office at the Healdsburg Surgery Center, I noticed that I'd missed a call from Dr. Weeks on my mobile phone. Still in my scrubs, I sat down in the chair in my office and returned his call. I hadn't expected to hear back from him so quickly, and I was hoping this was a good sign.

He answered right away. "I'm afraid I don't have an answer for you, but I suggest you get a book recently published by Dr. James P. Allen," he said. "He's the foremost authority on Egyptian medicine, who recently curated an exhibit for the Metropolitan Museum in New York. His book is *The Art of Medicine in Ancient Egypt*. It's the same title as the exhibit in New York that was held from September 2005 to January 2006."

I jotted down the title on a notepad while we talked, and I learned that the *Edwin Smith Papyrus*, written 3,600 years ago in its distinctive red and black ink, was one of the centerpieces of the exhibit. As soon as I hung up, I ordered a copy of Dr. Allen's book and then waited. How ironic that I might finally unlock the secret of hemayet in the United States after eight years of research, and three trips to Egypt with one phone call.

The book arrived several days later in the mail. I wasted no time opening the package and finding the page with Dr. Allen's translation of hemayet fruit in the "Recipe for Transforming an Old Man into a Youth." I found the recipe, and there it was. I finally had my answer.

Bitter almonds.

The instructions in the recipe suddenly made much more sense in Dr. James P. Allen's modern translation of the *Edwin Smith Papyrus*:

"One has to get a great many bitter almonds, comparable to 3 bushels. They have to be pulverized and put in the sunlight. After they

have dried completely, they have to be threshed like threshing grain. They have to be winnowed until only the kernals of them remain."

Botanically speaking, many of the nuts we eat are technically fruits. Walnuts, pecans, pistachios and almonds all fall into that category. That I knew. What I didn't know, however, was that almonds come in two forms—*prunus amygdalus dulces*, the sweet variety we consume and *prunus amygdalus amara*, better known as bitter almond.

I began my research into the pharmacology of bitter vs. sweet almonds. I used the Plant Pharmacopeia, which guided me to the answer. And since I had a background in organic chemistry, I was familiar with the formulas, which helped me to take the Egyptian text, paragraph by paragraph, and convert the English into an organic chemistry flow sheet.

I contacted Tom and Jules with the news right away, and suddenly the words "bitter almonds" rocketed back and forth as we burned up the phone lines and Internet connections from coast to coast. Jules followed up on it in the Merck Index, the industry's encyclopedia of chemicals, drugs and "biologicals," or medicinal preparations made from living organisms. First published by the Merck & Company pharmaceutical firm in 1889, the index has since grown to contain more than 10,000 explanations of known substances and chemical compounds. It is the professional chemist's bible.

As a chemist, Jules knew bitter almonds were poisonous because they contain cyanide. But what he didn't know was that bitter almonds also contain a substance known as mandelic acid.

"That's exciting," Jules said one day over the phone. "This has to be the active ingredient we are looking for."

"Yes, it is," I said, thinking, once again, that ancient Egyptian chemists were far more sophisticated than we realize.

Now that my search was over, there was the matter of finding them. I called Whole Foods Market first, but no one at the store had heard of them and I learned they were not sold at the store. After a little more research, I learned why.

Bitter almonds are poisonous. The bitter taste of these wild nuts comes from their naturally occurring cyanide, one of the deadliest toxins around. Historically, cyanide has been used by some clever— and not so clever—murderers to dispense with their victims. This is one of the main reasons why they aren't sold in the United States.

I was certain that bitter almonds were the way to go. Manniche notes in her book, *Egyptian Luxuries*, that in ancient times there were an abundance of bitter almonds in Sicily, just across from the Mediterranean from Egypt. The pieces were all starting to fall into place. But did bitter almonds still grow on this legendary island off Italy's coast?

I searched the Internet for answers and though I didn't find a direct answer, I did find recipes that called for bitter almond oil. Interestingly, all the recipes were for bitter almond cookies that originated from Turkey, Germany and Italy. Each recipe called for mostly sweet almonds with a small amount of bitter almond oil, which no doubt had the cyanide processed out of it. But I needed whole bitter almonds, not the oil for the "Recipe for Transforming an Old Man into a Youth." I soon found them available for sale in China, of all places. Given the history of this fruit in ancient Egypt, this took me by surprise, but the more I thought about it, the more it made sense. Like ancient Egyptian medicine, Eastern medicine has a long tradition of remedies based on plants that continues to this day. Evidently, bitter almonds are still used by herbal practitioners in China to relieve bronchial congestion. I also learned that apricot and peach pits are sometimes referred to as "bitter almonds" in China. Interestingly, the pits also contain cyanide (about 0.5 milligrams), which is considerably less than the 5.0 milligrams found in the average bitter almond.

I placed an order for a 20-pound sack of bitter almonds from a firm in the Shandong province that specializes in dried beans, nuts and fruits. And I was careful to specify that I wanted bitter almonds, not apricot or peach pits. The entire order set me back $175, plus $100 for shipping, and I requested the bag be marked "for experimental use only."

And then I waited. The order would take five months to arrive.

About Jules Zecchino

If you're a woman, chances are you've used a skin product touched by the hands of Jules Zecchino, a master of cosmetics chemistry. And even men's hands have been moisturized by Vaseline Intensive Care Lotion, which Jules formulated as a young chemist while working at Chesebrough-Ponds in the '80s. A huge success for the company, it also represented a prodigious start to a remarkable career.

Later at Avon, Jules headed a team to acquire an understanding about Asian women and their relationship to cosmetics, growing the company's sales in Japan from $80 million to $400 million during a five-year period in the '90s. And while working at Elizabeth Arden, Jules developed the company's signature Ceramide Capsules, which cleverly release a single application of skin-plumping serum.

But his longest tenure was at Estee Lauder, where he developed Fruition, Resilience, Idealist, Perfectionist and Illusionist skin renewal products. For the company's Clinique brand, he formulated the All About Eyes and Turnaround anti-aging creams. Along the way, Jules has been awarded five U.S. patents and for an unprecedented five years in a row, Estee Lauder skin products he developed won Marie Claire magazine's internationally-acclaimed Prix D'Excellence Beauty Awards. Twenty years later, when Jules left Estee Lauder in 2007, he was at the top of his game as one of the skin care industry's leading chemists, and went on to found Innovative Cosmetic Concepts, and then Skylar Brand, development firms for cosmetic formulations.

Chapter XVII

The Package Arrives
December 2008

T he package arrived on an ordinary day in December via UPS. To the delivery person, it was simply another parcel, albeit heavy and with a label printed in both Chinese and English. But after eight years of research and travel that unraveled a riddle possibly as old as the Sphinx itself, the package represented a watershed moment for me. I could have had the almonds delivered directly to Jules, but admit I was curious to see what they looked like firsthand.

The bitter almonds were contained in a brown burlap bag, and I wasted no time slitting open the top of it. About 11 of the small gray-brown nuts tumbled out, two spun in circles on the top of my desk. I watched them for a moment and when they stopped, I picked one up and licked it, curious to see if it tasted bitter. It didn't, but I wasn't about to take any chances by chewing it to see if that would release any of the bitterness. The smooth nuts were much smaller in size than sweet almonds and flatter than I expected, which meant their husks would probably have been an even greater headache to crack open by hand in ancient times.

Certainly, ancient Egyptians would have been right to "pulverize" and then winnow the husks away "like threshing grain" as the recipe directed. These bitter almonds came pre-shelled, which is how most consumers obviously preferred them, but I was after the chemical content of the nuts themselves and didn't need the husks.

Once the nuts were processed to remove their cyanide, I needed only to determine which effective ingredient or ingredients were left. And though I'd studied organic chemistry as part of my medical training, I needed a professional chemist to help me. I needed someone with broad experience in the cosmetics field, and open-minded enough to appreciate my crazy idea of resurrecting an ancient Egyptian recipe and formulating it into a modern product.

Fortunately, fate had previously arranged that for me, so I bundled up the bitter almonds and shipped them to Jules in New York City.

Formulating a new skin care product takes time and a great deal of experience, particularly when it involves a breakthrough ingredient. As an independent chemist, Jules maintains a private laboratory at a facility in Allendale, New Jersey. Once he received the package, Jules went to work analyzing them in his laboratory, closely following the Egyptian formula in the *Edwin Smith Surgical Papyrus*.

"It's absolutely essential to confirm that hemayet is in fact bitter almond," Jules said to me one day over the phone. "And the statement in the papyrus about it removing all signs of aging are intriguing to me, not to mention the comment that it is 'proven good a million times.' It's bold, yet proud. I can't wait to try it ourselves."

The instructions to keep washing the material in the river until the bitter taste was gone from the water exactly tracked the process Jules used in the lab as he dried and rinsed the ground almonds to remove the poisonous cyanide. However, Jules worked under a sophisticated chemical hood that shielded him from any toxic cyanide gassed off by the bitter almonds. When ancient Egyptians worked outdoors, we presumed they were working naturally in a well-ventilated environment.

But what about the unlucky person recruited to taste the water to see if any bitter cyanide remained? Jules and I wondered how many ancient Egyptians may have been sickened—or perhaps died—before chemical processing techniques were perfected. We also wondered what might have happened to any fish or wildlife living downstream from the runoff.

In any case, after Jules completed his work in the lab, following the ancient Egyptian recipe, the active ingredient we were left with was chemically consistent with the pure mandelic acid he had purchased commercially. Mandelic acid is an aromatic alpha hydroxy acid, a larger molecule than its cousin glycolic acid. A white crystalline solid, it is soluble in water and polar organic solvents. The name is derived from the German "Mandel" for "almond."

As for working with the ancient recipe, we learned that in the *Edwin Smith Surgical Papyrus*, translator Breasted speculated that the reason for the two boiling and drying steps is because "…the fruit taken was too large to be made up in one lot, and that the apothecary is therefore to divide it up into two lots merely for making." But what Jules discovered in the lab was that the second step in the recipe was to get rid of as much cyanide as possible, not because there was too much hemayet fruit.

At long last, we had strong scientific evidence that the hemayet fruit described in the "Recipe for Transforming an Old Man into a Youth" was indeed bitter almonds. And now we knew for certain that the key active ingredient in the recipe was mandelic acid, which we could purchase in its pure form. Jules disposed of the remaining bitter almonds, according to strict laboratory procedures.

Though Jules had 30 years of experience as a formulating chemist in the skin care industry, he mentioned he'd never worked with mandelic acid and what's more, the more research he did, he realized he was not alone.

"It had previously been used in some obscure ways, even in a few cosmetics products and some interesting properties had been attributed to it," Jules said one day over the phone. "However, it's clear from all these applications that the people currently formulating products with mandelic acid don't understand the underlying origins of the material. And I've also found a line of products featuring mandelic acid as a miracle cure for acne due to its known anti-bacterial properties. But what fascinates me is the fact that mandelic acid is an alpha hydroxy acid [AHA], so I believe it has a far greater potential."

"It sounds like we have something significant here," I said.

"I know we do. Based on my past studies with AHAs, I'm certain it will be effective."

But how effective? And could a serum concocted from this ancient ingredient stand up to modern day expectations of skin care?

What are AHAs?

Alpha hydroxy acids (AHAs) belong to a class of chemical compounds that include glycolic acid, one of the most popular ingredients in today's skin care treatments. Known primarily for their ability to speed the formation of new skin cells by exfoliating the old ones, AHA products first became widely available to consumers in the early 1990s.

According to Jules, the discovery of the action of AHAs on skin was an enormous breakthrough made in the '80s by his friend and fellow skin care research expert, Eugene Van Scott, M.D.

Says Jules, "I was fortunate to study with him when he was head of the dermatology department at Temple University in Philadelphia. Gene was a huge proponent of AHAs. He did many pioneering clinical studies on the materials, focusing mainly on glycolic acids, and had the foresight to patent the use of them in cosmetics. In fact, I created one of the first AHA products for the industry—Fruition by Estee Lauder—based on his work."

Chapter XVIII

Creating a Serum

O nce Jules crushed his first batch of bitter almonds, it took him nine months to develop the first samples of serum. As each new ingredient was added to the formula, the mixture had to be carefully monitored to make sure the properties of one material didn't lessen or negate the effectiveness of the others. Fortunately, Jules had an extensive amount of experience.

From his years in the industry, Jules knew he wanted to include other ingredients that would augment mandelic acid. In what we decided would be a new, non-prescription serum for the skin, he included the potent anti-oxidant resveratrol, glucosamine for skin-cell turnover, as well as Boswellia (or frankincense), an anti-inflammatory since we believe aging is an inflammatory process, licorice root and other agents that facilitate penetration through the stratum corneum (the outermost layer of the skin).

Unlike moisturizers, which typically form an outer barrier that helps prevent the skin from drying, serums are formulated to deliver their active ingredients into deeper layers of the skin.

Jules had a plan to use them at specified levels for maximum efficacy and safety. The mandelic acid was used at the best level for the serum as determined by its solubility in the formula, but it took

Why Reformulate the 'Face Paste?'

Though ancient Egyptians had apparently perfected their elixir, which translator James Henry Breasted referred to as the "face paste," the original formula needed an update for two reasons.

Safety. A product must have the correct pH level to maintain the naturally acidic pH barrier of the skin because when a product disrupts the natural pH level, your skin is prone to damage and infection. Like most acids in their pure form, mandelic acid is too concentrated for direct application to the skin and must be blended with other substances to achieve an overall safe pH level. In addition, during the testing process, Jules found that the original formula was prone to residual separation of the ingredients and premature spoilage, another safety hazard.

User-friendliness. The original formula truly was a face paste. In fact, it had the same consistency of old-fashioned peanut butter, where the oil must be poured off before using it. This is not the sort of quality consumers want, no matter how spectacular the end results. As for mandelic acid, in its commercial form, it is highly sticky and not easy to use.

countless hours in the lab to hone in on the optimum formula for modern-day use.

This is where substances from Grant Industries came to the rescue. After consulting with Tom, Jules chose the company's Gransil DM5T, a silicone elastomer gel to add to our serum. As the name suggests, elastomers add an "elastic" quality to a formulation, allowing it to smooth more easily over the skin.

"Gransil DM5T works like a charm, greatly improving the formulation's 'slip,' spreadability and overall silky afterfeel," Jules explained one day over the phone. What's more, as a result of its chemical and physical structure, this high-end elastomer would also gently refract light off the face, visually reducing the appearance of surface lines and wrinkles—a bonus for an antiaging product. After a few more adjustments, Jules had developed the base formula to his satisfaction and shipped preliminary samples of the new serum, so to see what I thought of its smell and feel. I already knew I'd want to conduct a clinical trial to test its effectiveness on people's skin once we had agreed on a final version of the product.

We had yet to give the formula a name, but in less than a year, we had our first modern reformulation of the recipe inspired by ancient Egyptians. But would the formula work?

What is an Anti-Aging Product?

If you look online, you'll find that anti-aging creams are predominantly moisturizer-based cosmeceutical skin-care products, marketed with the promise of making the consumers look younger or preventing the signs of aging. The term "anti-aging" has been around for a long while. It's mostly a marketing term, but it implies that a given product will fight aging. Generally, if a product reduces wrinkles, it's called anti-aging.

If a product falls into the category of anti-aging, it should contain an active ingredient. One of the active ingredients in our products is mandelic acid.

Chapter XIX

Measurable Results

Long before Jules delivered the final version of our serum, we knew we wanted scientifically verified results, and this was a top priority. Having used surgery, lasers, fillers, Botox and in-office skin peels to deliver results my patients can see and feel, I wanted a serum that could also deliver results and be scientifically measured.

An implant, for example, can extend a chin by as many as seven to eight millimeters, lending a more balanced look to your profile. When performing a Serial Vector lift (my version of a facelift), I stitch connective tissue fascia tissue by up to five centimeters along several points closest to the jowls to tighten the laxity brought about by age. And a filler such as Juvederm can add several millimeters of plumpness beneath the skin to minimize nasolabial lines (the smile lines that flank your nose).

Since patients come to me to see the dramatic results these kinds of procedures can offer, I wanted nothing less from any other product I would be endorsing, especially one promoting improvements to the skin. Traditionally, products claiming improvement to the skin is an area of study that has largely relied on subjective means for analysis. To measure the effectiveness of skin care products, dermatologists and the skin care industry typically use visual evidence from before-

and after-photos and reports from study participants about how the skin feels, improvements that are evaluated by the senses only. (Next time you're flipping through a women's magazine, look for a skin care advertisement and the footnote often says something like this: 90 percent of the women who used this product for six weeks said they "felt" they looked 10 years younger.)

But no matter how careful researchers are when taking photographs, the results are prone to variances in lighting, angles, distance, posture, the subjects head tilt and so on. And what "feels" smoother to one person, may not to another.

We were looking for scientifically measurable results, but at the time there was no device that could measure this available in the United States. I did some research and found a Cutometer, manufactured in Germany by Courage + Khazaka, a firm that specializes in skin-testing equipment.

The Cutometer was designed to measure the "visco-elasticity" of the skin. Skin is viscous, existing as neither a solid nor a liquid. We had come to understand that skin is more like a living form of fabric with elastic properties.

The Cutometer arrived in February of 2009. The seven-pound device came with a large binder of instructions and practical guidelines that weighed almost as much. Once I learned the basics, I tested myself. I connected my laptop to the Cutometer, placed the tip of the wand with a tiny hole in it against my lower eyelid (the most wrinkled area on my face) and felt a silent and almost imperceptible tug on my skin for a few seconds before it released. (Inside the wand is a spring-loaded device with a light sensor that measures the elastic movement of the skin up and down.) The process took about 10 seconds and produced a readout to measure the elasticity of my skin. My lower eyelid elasticity ratio measured 38 percent. (In contrast, I later measured the elasticity of my then-teenage sons' lower eyelid area and they both tested over 90 percent. When I tested a 65-year-old female smoker, her eyelid elasticity was 23 percent.)

Now that I had a baseline for the skin below my lower eyelid (38 percent elasticity), I began to apply the serum on my face twice a day, paying particular attention to the wrinkly skin on my lower eyelids.

I continued the routine religiously for a month, and then took an "after" measurement with the Cutometer. I admit, as much as I'd hoped the serum would work, even I was taken by surprise by the dramatic results. After applying the serum on my face and lower eyelid area, the elasticity had increased to an incredible 88 percent! I repeated the test on the other eyelid and the readout was above 85 percent. The elasticity was almost as high as my teenage sons' baseline readings. I don't recall whether I called Joni or Jules first with the news, but if I'd misdialed and reached Mary's Pizza Shack, I would have shared the news. The results were extraordinary, to say the least. I came to realize later, that an objective increase of 5 to 10 percent is significant.

But I knew from my experience as a clinical researcher, that I'd need to stop using the serum for a month and take another measurement. (This type of testing is performed on many medications to demonstrate that it's truly the drug causing the effect, not happenstance.) After going without the serum for one month, the elasticity on my lower eyelid dropped back to about 35 percent. Then when I returned to my twice-per-day treatment schedule for a month and tested again, the elasticity was back and above 80 percent.

Amazing.

And this was the defining moment that I suspected we were onto something great, but the test I'd conducted on myself represented what those in the industry would consider one "data point." We would need more data points before I could validate my results to the medical community. The next step was to conduct a clinical study.

Chapter XX

Clinical Studies
2009 - 2010

F ortunately, I had a built-in pool of potential participants through my practice in the San Francisco North Bay area. Patients coming to see me were already interested in improving their appearance. Whether it was wrinkles, sun damage or other problems related to aging, they wanted a fix. But my patients had to be willing to wash with Ivory Soap for one week before starting with the serum. This is known as the "wash-out period."

What's more, to maintain consistency and to avoid the possible effects of other skin products, participants were instructed to forgo any make-up, moisturizers, toners and sunscreens for five weeks. (Naturally, not everyone I attempted to recruit could meet these requirements due to job expectations, sun sensitivity or other practical reasons.) Nevertheless, I recruited 26 patients (24 women and two men), ranging in age from 42 to 68.

At the start of the study, I used the Cutometer to measure one lower eyelid, cheek, jowl and neck point on each participant to get baseline elasticity percentages. Patients were randomized for "sidedness," which meant I tested either their left or right side consistently throughout the study. Like right- or left-handedness, people unconsciously make different expressions, sleep with one side of their cheek against a pillow and get different sun exposure on each side of their face, so studies must account for this. I also took photographs of each person to provide visual evidence that might correlate with any changes in elasticity measured by the Cutometer. Next, study participants began using the serum and a night crème we developed with similar ingredients to the serum but with moisturizing agents such as Poly Glutamic Acid (known to be four times more hydrating than hyaluronic acid), twice a day and returned each week for a month for measurements and photographs. The Cutometer recorded amazing elasticity gains after only one week.

But what we found at the conclusion of the study was surprising to everyone. On average, participants saw an increase in lower eyelid elasticity of a remarkable 26 percent. But perhaps most interesting, I noticed two distinct groups emerge based on their starting elasticity before the clinical study. Those whose skin elasticity was characterized as "poor," meaning under 50 percent elasticity, saw an increase in elasticity of 47 percent. And those participants whose skin was characterized as "good," meaning more than 50 percent elasticity, had an increase of 26 percent.

What does this mean? The results suggest that elasticity and viscosity are keys to great skin. Viscosity is the firmness of your skin and how far away it can be pulled from your body. The farther you can pull away your skin, the less viscous it is. The skin of a child or younger adult, for example, is firm and harder to pull away, which means it has good viscosity. The ingredients in the serum and moisturizer trigger both collagen and elastin production, which improves elasticity and viscosity.

Skin Elasticity Study

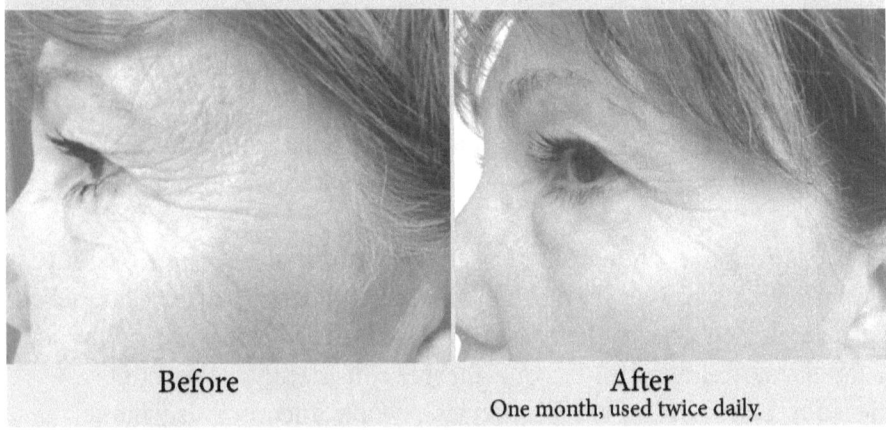

Before   After
                              One month, used twice daily.

The results suggest that human skin has a spectrum of visco-elastic properties. Skin with poor starting elasticity must first experience gains in its elastic properties before its firmness can increase. For example, my then

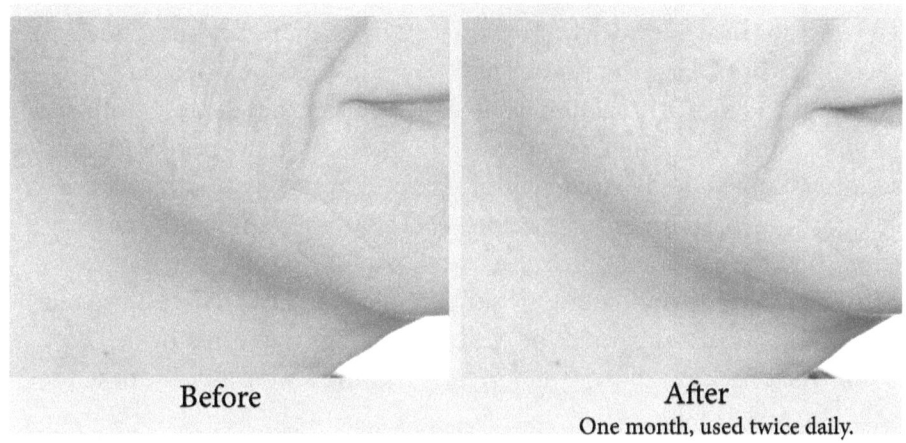

Before

After
One month, used twice daily.

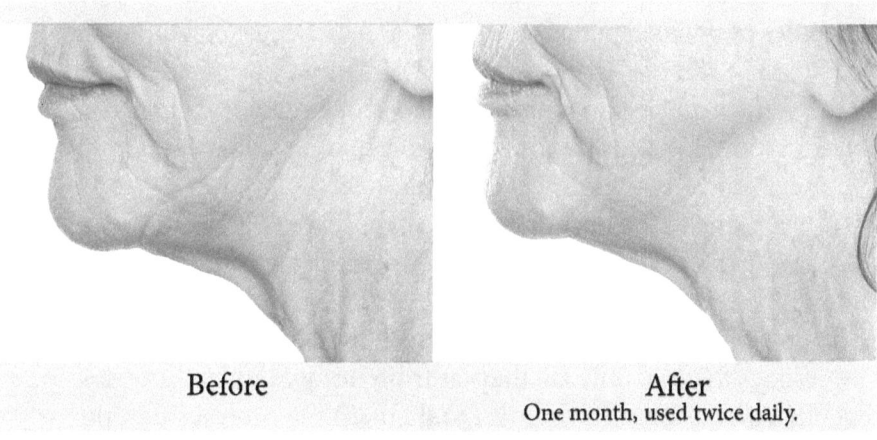

Before

After
One month, used twice daily.

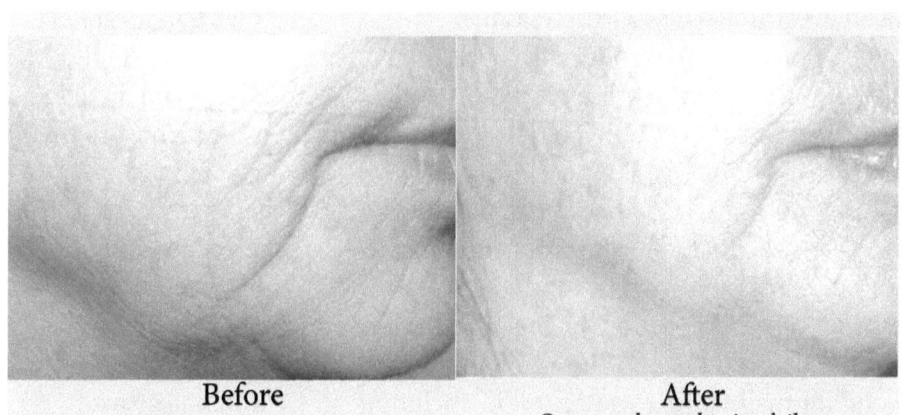

Before

After
One month, used twice daily.

teenaged sons with elasticity ratios above 90 percent have almost nowhere to go in terms of elasticity gains. They would instead see a decrease in how far the Cutometer suction wand could pull their skin away. Think of a rubber band being folded on itself. It does not become more elastic. It becomes firmer—and firm, youthful skin displays fewer wrinkles.

In sorting through the before-and-after photographs, I was truly astonished. Not putting much stock in this subjective measurement technique (or in my photography skills), I could still clearly see dramatic improvements that mirrored the elasticity gains as measured by the Cutometer. Seeking an explanation for why our new products worked so well, I turned to Jules.

One of the secrets could be found in the molecular structure of mandelic acid, he said. "During my early work with Dr. Eugene Scott at Temple University, we discovered that bigger alpha hydroxy acid [AHA] molecules performed better on the skin and caused less irritation than the smaller AHA molecules. So as a large AHA molecule, mandelic acid acts on the surface of the skin longer and does not irritate at higher levels of concentration, like most of the shorter-chain molecules normally found in treatment products today such as glycolic acid."

Experts in the field of skin care believe that because smaller molecules penetrate the skin faster, they're more likely to cause redness and skin irritation in some people. In addition, glycolic acid and retinoic acid (a metabolite of vitamin A) have been used so much over the past few decades that not only are they far from unique, and it is possible that our skin has become immune to them. In the clinical study, no one reported experiencing any skin irritation with our mandelic acid products. And because mandelic acid is a larger AHA molecule that has been underutilized as a skin treatment, we believed it made an ideal next generation skin care product. To optimize its use in our products, Jules combined it with glucosamine, which works directly with mandelic acid to make the skin smoother and firmer, by speeding up the turnover of fibroblasts (skin cells).

Jules also added powerful peptides and growth factors to stimulate collagen, elastin and hyaluronic acid synthesis deep within the skin to decrease lines and wrinkles and improve firmness. And to provide the best wrinkle-reduction action, he included a whey protein complex derived from milk. Said Jules, "Whey proteins have many interesting molecules good for maintaining youthful skin, including tgf-beta [tissue growth factor beta], which directly stimulates fibroblasts in the skin to produce collagen. The particular fraction we use out-performs all the other peptides specifically designed for this purpose. And because it is a growth factor, it also stimulates elastin and hyaluronic acid synthesis, the main materials in the skin responsible for keeping it firm and elastic."

Jules was instrumental in getting commercial production started at Englew ood Lab in Englewood, New Jersey. With an initial output of 5,000 units under way, we worked to launch sales of our serum in my offices in San Francisco and Healdsburg and online through Amazon.com, Inc.

Independent clinical study

In 2010, we approached the televised home shopping market. But one well-known home shopping network uses independent laboratories to test the products before marketing them on their network. For skin care, this means a product must produce results as advertised. As a result, Essex Testing Clinic, Inc. conducted an independent clinical study of the serum, which we'd officially named *Stanley Jacobs MD Visco-Elastic Transforming Serum with Mandelic Acid*. As a nod to the past, I wanted to incorporate the word "Transforming," which is the same term used in the ancient recipe.

The independent study by Essex included 30 female participants over an eight-week timeframe. The participants began with an elasticity ratio of 52 percent (in the "good" range) on their cheeks and experienced an average increase of 16 percent, which tracked closely with the same range I measured with the Cutometer in my own study. This time around the night crème was not part of the study; only the serum was used. The area to be tested would be the cheek area, and we were delighted to learn that Essex would also measure results with a Cutometer. What's more, while they were testing for elasticity they would simultaneously conduct other types of skin tests.

The results from the independent study were even more remarkable than the results from my study. Here are some of the highlights.

• Skin elasticity increased in 97 percent of subjects.
• Skin moisture was significantly improved 15 minutes after a single application;

• Skin barrier function (trans-epidermal water loss) was significantly improved after two, four and six weeks of product use;

• Skin firmness was significantly improved after two, four and six weeks of product use;

• Global fine lines/wrinkles were significantly improved after two, four and six weeks;

• Skin evenness/skin clarity was significantly improved after six weeks of product use; and

• Age spots/skin discolorations were significantly improved after two, four and six weeks of product use. (This was calculated by using a computer gray scale.)

In addition, the serum was met with a high-level of consumer acceptance, according to the report from Essex. All the participants felt the product was gentle and non-irritating and 93 percent of the subjects reported an improvement in the skin's overall appearance and the participants would recommend the product to a friend. What's more, 90 percent of the participants reported improved skin hydration and skin tightness/firmness and would purchase the product. There was no irritation observed during the course of the study.

Naturally, we wondered how our serum compared against similar skin care products. Without disclosing the names of other products, Essex gave us a ballpark figure for the products they'd tested over the years that contained retinoic acid an ingredient commonly used to eliminate wrinkles. The average gain in skin elasticity of the cheek as measured by the Cutometer continues to range from about 20 to 50 percent. In addition, Essex's lead chemist told us our serum fared better than most products their lab had tested with the Cutometer. We also learned that more than a few skin care products fail to meet the Home Shopping Network's test standards, and as a result are never sold by them.

With our own clinical results and the report from Essex Testing Clinic, we were on our way to revolutionizing skin care with an ancient formula.

Chapter XXI

An Ancient Treasure

A s I look back at the incredible journey that began in 2000 and took nearly a decade of research to unlock the mystery of the hemayet fruit, I can't help but marvel over the innovative genius of ancient Egyptians. How did they come up with the formula to transform an old man into a youth? Why did they use bitter almonds and not some other substance? Did they *intend* to develop an unguent that would improve wrinkles from the skin? I use the word "intend" because an important clue to what might have been the original use of the recipe was written in the instructions for applying the unguent:

> *"Anoint a man with it. It is something that repels a cold from the head. If the body is wiped with it, what results is rejuvenation of the skin and repelling of wrinkles, any age spots, any signs of old age, and any fever that may be in the body."*

Given the level of sophistication they reached with their medicine and surgery, it's no surprise ancient Egyptians were experimenting with fever-reducing poultices that could be externally applied to the head and body. (Nowadays, ice packs largely play that role.) Research by Lise Manniche, who wrote *An Ancient Egyptian Herbal*, shows that bitter almond oil was a key ingredient in "Metopian," an ointment developed by the Egyptians to "open the vessels" and produce sweat, which helps cool the body. Perhaps it was similar to what we know as Vick's VapoRub today. But clearly, there was something in bitter almonds the ancients considered key to treat fever. If this was the case, then wrinkle reduction might have been discovered as a by-product of the poultice. Without ice or antibiotics, a fever in 1500 BC could easily rage in a person for a week or longer. If the bitter almond poultice with its powerful dose of mandelic acid was applied at least once a day to a patient's face, improvement to the skin would likely have been noticed in as little as a week, similar to the clinical study and the one conducted by Essex Testing Clinic. Whether ancient Egyptian patients survived the fever with the bitter almond poultice, chances are they had great-looking skin, even if only for the afterlife.

The continuing mystery of hemayet

During this odyssey, I often wished I could be transported back in time to witness the ancient Egyptians process of discovery. Though we eventually deciphered the meaning of "hemayet " to create a serum that's scientifically proven, the mystery of their process remains. The royal chemists undoubtedly spent many hours, months and years developing their formulas. And if they happened upon the "Recipe for Transforming an Old Man into a Youth" by accident, the intent was undeniably there to pursue, perfect and record the resulting formula for improving the appearance of the skin. As the *Edwin Smith Surgical Papyrus* states, the recipe was "proven good a million times." But regardless of how it happened, it still amazes me to think that 3,600 years ago, the greatest civilization on earth was working on a skin care product for queens and pharaohs.

But how did the royal chemists know the process they followed would create an unguent that would make the skin tighter? How did they know to follow a certain sequence—to husk and winnow, sift, cook and wash and to follow the process several times? Clearly it was trial and error. They had to have done it many, many times.

Ancient Egyptians were brilliant at medicine and surgery. When I found the formula in Breasted's book, it didn't surprise me that they took the time to write it down. Their world was different. They didn't have to bother with marketing or advertising campaigns, and they weren't concerned with being sued. In ancient times, Egyptian chemists had nothing to gain; the only person they had to impress was the pharaoh and they wouldn't have taken the time to write down their formula for more youthful-looking skin if it didn't work.

The original *Edwin Smith Surgical Papyrus*

During the autumn of 2010, I was invited to present my research on the clinical study at the annual conference of the American Academy of Facial Plastic and Reconstructive Surgery in Boston, Mass. The event was attended by more than 500 of my counterparts from the United States and around the world. The purpose of the conference was to keep up-to-date with new developments in the industry and share ideas. To have my paper accepted for presentation was an exciting honor.

    While attending the conference I informally polled surgeons through a show of hands and found that no one had ever heard of the Cutometer. I realized that physicians were not accustomed to objectively measuring skin quality before and after treatment. Currently, we all perform different procedures, and comparing results is difficult. However, if every physician used a device that allowed the outcomes of surgeries, lasers, peels and creams to be compared equally, it would hold doctors accountable and give patients the data to better select which procedure would work best for them. The Cutometer is one way in which physicians can rise above the varying tide of subjectivity and believe in true and honest data. It's my hope that all doctors whose practice involves skin care start using this instrument to advance the various methods of treatment. Meanwhile, the Cutometer is an important diagnostic tool for my practice and part of my approach to skin care today.

    My associate, Eric Culbertson, M.D., and I documented our research in a paper titled, "Effects of Topical Mandelic Acid Treatment on Facial Skin Viscoelasticity." (See Appendix on page 154 for abstract of article.) In September 2018, I presented my research at the European Academy of Facial Plastic Surgery in Regensburg, Germany. As of this writing, our research was accepted by the journal, *Facial Plastic Surgery,* and is scheduled for publication in 2019. My research exposes the surgical community to the concept of skin measurements, and hopefully in the end the patients will reap the benefits. After all, isn't that the whole idea?

During that trip to the conference of 2010, I travelled from Boston to New York City where Jules and I had the opportunity to see a portion of the original *Edwin Smith Papyrus* by special request. Though the 3,600-year-old medical document is owned by the New York Academy of Medicine, it was on loan at the time to the New York Metropolitan Museum of Art, where it was undergoing restoration. On a bright September morning at 11 o'clock, Jules and I were led through a corridor in the museum's Egyptian exhibit past a number of Fayum sarcophagi with encaustic hot wax paintings on the wood coffins, depicting the person buried inside. Their faces had lifelike eyes that appeared to follow us as we turned through corridor after corridor and entered a locked room filled with ancient busts and books stacked on shelves along the walls. The room was partially divided into two areas. Two other people were already inside speaking quietly to one another, a man and woman working on a project in the space to our left.

　　While we waited for the papyrus to arrive, I relayed the story of our serum to the curator. Soon, an elderly gentleman entered the room pushing a trolley with an ordinary looking flat cardboard box. The curator stepped forward and carefully opened the lid, and there it was—a section of the *Edwin Smith Surgical Papyrus*. The hieratic writing was jet black and appeared almost fresh. Some of the ink was a deep blood-red color with fresh brush strokes clearly visible at the end of each word. The document was neat and tidy, with most of its beige surface taken up with writing. It gleamed back at us with such vibrancy that it seemed as though the talented scribe had simply set his quill down for a few fleeting moments before returning to continue his work. I own 30-year-old medical texts that are much more faded. The ancient papyrus was magnificent. After nearly a decade of research to unlock the mystery of hemayet, it was a rare privilege to see the original work of art and science before us.

　　Developing the serum from this ancient document was the result of a string of synchronistic events that seemed to randomly present themselves, leading us to bring this ancient formula to the modern-day world of skin care. Much to my surprise, fate was about to deliver another hidden treasure. Does coincidence really exist? To me it seemed to be a pre-ordained pathway that was meant to be.

After viewing the document, I finished telling the curator our story of recreating the serum and mentioned that I needed to thank Dr. James Allen, who had ultimately resolved the mystery of hemayet and directed "The Art of Medicine in Ancient Egypt" exhibit for the Met in 2005. The woman we'd seen when we first arrived turned to me.

"I'm Dr. Allen's wife," she said, pausing to smile. "I couldn't help but overhear your story. Fascinating."

We chatted for a while and I offered to send her some of our serum. She gave me her address in Rhode Island, and I learned that Mrs. Allen visits the museum every three months. I smiled at this unexpected "chance" encounter.

Since that day, I've struck a friendship with Mrs. Allen and her husband, Dr. James Allen. Over the years, they've encouraged me to write this story and present at an ARCE meeting some year. The fortunate stroke of serendipity that led to the rediscovery of this serum is not lost on the Allens. Egyptologists had never tried to decipher the mystery of hemayet since they're not necessarily interested in skin care, and most plastic surgeons aren't going to take the time to read a papyrus about ancient Egyptian surgical techniques. But this plastic surgeon and now Egyptologist is interested in both mysteries and helping humankind stay healthy and youthful-looking through their lives.

In retrospect, I sometimes wonder—in our pursuit for youthful skin—why modern-day cosmetic chemists had never investigated the practice of ancient Egyptian skin care. The Egyptians were geniuses when it came to the preservation of the dead, so it only stands to reason that the pharaohs would take as great of care in their grooming and skin care. Nevertheless, modern scientists didn't fully understand that ancient Egyptians were cosmetic scientists, who were surprisingly sophisticated in their practice of skin care. This serum is an homage to them. What better way to give them tribute, than to resurrect a molecule in skin care that they crafted thousands of years ago, and bring it to life to help people across the globe today.

Cleopatra: Fact & Fiction

So much has been written about Cleopatra, the legendary last queen of Egypt, that fact and fiction have become all but indistinguishable. Did Cleopatra really hide in a rolled-up rug and have herself delivered to Julius Caesar? (Truth: She was in a large cloth sack.) Did she truly drink a pearl dissolved in wine at a banquet to impress Mark Antony? (It is scientifically possible.)

The legendary leader was said to have been fluent in nine languages, easily able to compute math in her head and possessed a good sense of humor. Legend has it that she kept her skin soft and beautiful by regularly bathing in milk and honey. Throughout history, much has been written about her beauty secrets and the creams she used to keep her skin flawless. Being the savvy woman that she was, Cleopatra surely made the most of her physical assets by using the finest beauty products available to her.

Might she have used the ancient Egyptian skin unguent produced by the "Recipe for Transforming an Old Man into a Youth?" It's a strong possibility. A substance valuable enough to be placed in a container of costly stone wouldn't be available to the everyday Egyptian. Medical knowledge and formulas were respectfully passed from one generation of doctors and priests to the next, so as a pharaoh, Cleopatra would have had access to them.

Chapter XXII

Launching the Serum

The night our serum was officially launched was an unusually clear December evening in San Francisco in 2009 at the de Young Museum in Golden Gate Park, where the touring exhibit, "Tutankhamen and the Golden Age of the Pharaohs" was held several months earlier. It was windy and chilly enough that Joni and I longed for the warmth of Egypt.

The party was held on the top floor of the museum's new modernist tower, which features 360-degree views of the city with floor-to-ceiling windows. About 100 guests joined us, and Jules and Tom flew in from New York to celebrate a night filled with food, music and fun. In honor of the occasion, I wore a white shirt with Egyptian motifs under my tuxedo, and Joni's Eye of Horus pendant graced the neckline of her one-shouldered gown, reminiscent of the Greco-Roman period. I saw more than a few women who exaggerated their eyeliner in a nod to the ancient past that inspired our present-day formula. It was truly a celebration after nearly a decade's journey. Guests left with a bottle of serum in the now signature lapis lazuli blue and gold container, which includes the profile of Nefertiti in gold.

The eternal beauty of Queen Nefertiti
Though Cleopatra is often thought of when it comes to beauty

routines of ancient Egyptian women, Nefertiti was by all accounts the more beautiful. An ancient sculptor captured her graceful features for eternity on a 17-inch-tall bust of such astonishing beauty that it became even more famous than her name. To this day, many facial plastic surgeons use a copy of Nefertiti's three-dimensional bust in their practices as a reference, including me.

The life of Queen Nefertiti remains a mystery today. Nefertiti was the great royal wife of Pharaoh Akhenaten. Though she is believed to be the stepmother of the King Tutankhamen, after studying the composition of her face and King Tut's mummy, I've been convinced for years that Nefertiti was his mother. I've often wished I could mask King Tut so I could reconstruct what he looked like when he was alive and compare it to the bust of Nefertiti. During the early part of 2015, headlines brought attention to Nefertiti once again and Egyptologists now suspect she may very well be his mother after all.

Nefertiti's connection to King Tut and the whereabouts of her remains may forever remain a mystery. Excavations in 2016 show that her tomb may be resting directly behind King Tut's tomb. Is King Tut the son of Nefertiti? Perhaps we'll never know, but I suspect there's a strong possibility. Her image aptly graces the bottle of our products, and it is highly probable that she once used the elixir recorded on the 3,600-year-old papyrus. Medical knowledge and formulas were respectfully passed from one generation of doctors and priests to the next, so the pharaoh and queen would have access to them.

After the launch party that evening, my family and I were making our way through the parking lot when I happened to look up at the night sky and caught a glimpse of a single star peeking through a fleeting gap in the now foggy sky. I couldn't help but think about my first evening in Cairo when I gazed up at the night sky.

I wondered what the ancient Egyptians would have thought of this journey I traveled, and the effort it took to reformulate their elixir. Egyptian culture is steeped in preservation and making the skin appear smooth and silky. When I first came across the recipe in Breasted's book, I thought it would be simple. In reality, it was an elaborate formula that was well thought out and researched. The ancient Egyptians were master chemists and physicians, and they had to have worked with this formula hundreds of times to get it just right. They were geniuses. Somehow, I think these remarkable people knew all along the eternal secret to great skin care.

"Nefertiti" means beauty has come, and each time someone uses our serum, the past and present intersect. This journey remains one of the high points of my life, bringing the wisdom of ancient Egyptian chemistry and modern-day science together in an exciting new way to revolutionize skin care today.

Tutankhamen's Missing Oils & Unguents

Tomb robbing in ancient Egypt is as old as the tomb's themselves. Having detected a re-plastered hole in the entrance of Tut's tomb in 1922, archeologist Howard Carter realized he was not the first to have gone inside. "The tomb then was not absolutely intact as we had hoped," he later wrote in *The Discovery of Tutankhamun's Tomb*. "Plunderers had entered it, and entered it more than once.

From a record made at the time of Tutankhamen's funeral, Carter estimated about 60 percent of the jewelry from inside the tomb was missing along with a series of metal vessels. Forgotten or abandoned in haste by the looters were seven solid gold rings wrapped in a robber's handkerchief and stashed in a box in the tomb's annex. Uninterested in the food provided for Tut's journey to the afterlife, the thieves left it behind as well, so among Carter's discoveries was a small red pottery jar containing 30 almonds. But were they bitter almonds or sweet? In *The Complete Tutankhamen*, author Nicholas Reeves describes them as "nibbles" of the *Prunu dulcis* (sweet) variety, and author Lise Manniche writes that the almonds were in a jar labeled "wnt," as opposed to "hemayet." Still, these almonds were never chemically tested to confirm they were indeed sweet almonds, or the bitter variety.

In addition, Tut's tomb included jars and vessels and some of them were carved from alabaster and were amazingly elaborate. The most magnificent stood 28 inches tall, its center vessel flanked by two statues of the Nile god, each holding a lotus flower. If ever there were a "vase of costly stone," as mentioned in the "Recipe for Transforming an Old Man into a Youth" to store the elixir, this was it. The formula had been recorded in the *Edwin Smith Papyrus* about 200 years before Tut's reign (1341 to 1323 BC). It is well within the realm of possibility that our ancient hemayet preparation was stored in one of the jars or vessels in Tut's tomb. This elaborate royal vessel, however, was intended for perfume as it contained the residue of an aromatic gum resin mixed with fat.

In 2004 a group of Spanish scientists used liquid chromatography and mass spectrometry to analyze wine residue found in several clay jugs buried with Tutankhamen. They confirmed the presence of malvidin-glucoside, the compound in red grapes that gives red wines their color. Two years later, this same group of scientists analyzed more jugs found with the young king and discovered the first-ever chemical evidence of white wine in ancient Egypt.

More of these scientific studies will undoubtedly continue. In addition to Tut's alabaster containers, many other jars that once held ancient cosmetics have been recovered in Egypt and reside in museums worldwide. With the chemical composition of the "Recipe for Transforming an Old Man into a Youth" now known, it would be relatively easy to compare the residual contents in these vases of "costly stone."

Bibliography

Agache, Pierre and Humbert, Philippe. *Measuring the Skin.* Springer-Verlag, Berlin, Germany, 2004.

Allen, James P. *The Art of Medicine in Ancient Egypt.* The Metropolitan Museum of Art, New York, 2005.

Belzoni, Giovanni. *Narrative of the Operation of Recent Discoveries in Egypt and Nubia.* John Murray: London, 1820. http://www.archive.org/details/recentdiscovpyra00belz

Bettmann, Otto L. *A Pictorial History of Medicine.* Charles C. Thomas: Publisher, 1956.

Breasted, James Henry. *Ancient Records of Egypt,* Vol. 1. University of Illinois Press, 1906 (and 2001).

Breasted, James Henry. *The Edwin Smith Surgical Papyrus.* The University of Chicago Press, 1930.

Bryan, Cyril P. *The Papyrus Ebers* – Translated from the German Version. London: G. Bles, 1930. (excerpts and page citations found at: http://www.dinweb.org/dinweb/DINMuseum/Ebers%20Papyrus.asp) http://www.dinweb.org/dinweb/DINMuseum/Ebers Papyrus.asp
Clayton, Peter A. Chronicle of the Pharaohs. Thames and Hudson, Ltd., London, 1994.

Diamond, Jared. *Guns, Germs, and Steel – The Fates of Human Societies.* W.W. Norton & Company, Inc., New York, 1999.

Ebeid, Nabil I. *Egyptian Medicine in the Days of the Pharaohs.* General Egyptian Book Organization: Press, 2000.

Emsley, John. *Molecules of Murder – Criminal Molecules and Classic Cases.* The Royal Society of Chemistry, Cambridge, 2008.

Evans, Helen C. and White, Bruce. *Saint Catherine's Monastery, Sinai, Egypt – A Photographic Essay.* The Metropolitan Museum of Art, New York, 2004. Fagan, Brian. *The Rape of the Nile – Tomb Robbers, Tourists, and Archaeologists in Egypt.* Westview Press, A Member of the Perseus Books Group, 2004.

Filer, Joyce. *Egyptian Bookshelf: Disease.* University of Texas Press, Austin, 1996.

Fluckiger, F.A. and Hanbury, D. Pharmacographia – *A History of the Principal Drugs of Vegetable Origin.* R.P. Singh: International Book Distributors, India, 1986.

Gawande, Atul. Better – *A Surgeon's Notes on Performance.* Metropolitan Books, Henry Holt and Company, New York, 2007.

Hawass, Zahi. Bibliotheca Alexandrina – *The Archeology Museum. The Supreme Council of Antiquities,* Cairo, 2002.

Hurry, Jamieson B. Imhotep – *The Vizier and Physician of King Zoser.* Oxford University Press, 1926.

Ikram, Salima and Dodson, Aidan. T*he Mummy in Ancient Egypt – Equipping the Dead for Eternity.* The American University in Cairo Press, 1998.

Jacobs, Stanley W., BSc, MSc, MD, FRCS and Culbertson, Eric J. Culbertson, MD. *Effects of Topical Mandelic Acid Treatment on Facial Skin Viscoelasticity.* Facial Plastic Surgery, December 4, 2018. Thieme Medical, New York.

Klein, Carl H. von. *The Medical Features of the Papyrus Ebers.* Chicago: Press of the American Medical Association, 1905. http://www.archive.org/details/cu31924000900849

Koch, R.J. and Cheng, E.T. "Quantification of Skin Elasticity Changes Associated with Pulsed Carbon Dioxide Laser Skin Resurfacing."

Archives of Facial Plastic Surgery, Vol. 1, No. 4, 1999. http://archfaci. ama-assn.org/content/1/4/272.full

Lucas, A. and Harris, J.R. *Ancient Egyptian Materials and Industries.* Dover Publications, Inc., Mineola, New York, 1999.
Manniche, Lise. *An Ancient Egyptian Herbal.* University of Texas Press, Austin, 1989.

Manniche, Lise. *Egyptian Luxuries – Fragrance, Aromatherapy, and Cosmetics in Pharaohnic Times.* The American University in Cairo Press, 1999.

MacLeod, Roy. *The Library of Alexandria – Centre of Learning in the Ancient World.* The American University in Cairo Press, 2002.

Nunn, John F. *Ancient Egyptian Medicine.* University of Oklahoma Press: Norman, 1996.

Plutarch. *Isis and Osiris.* Loeb Classical Library edition, 1936. http:// penelope.uchicago.edu/Thayer/E/Roman/Texts/Plutarch/Moralia/ Isis_and_Osiris*/E.html#T383

Reeves, Nicholas. *The Complete Tutankhamun.* Thames & Hudson, Ltd., London, 1990.

Schiff, Stacy. *Cleopatra: A Life.* Little, Brown and Company, New York, 2010.

Smith, Craig B. *How the Great Pyramid Was Built.* Smithsonian Books, HarperCollins Publishers, 2004.

Tiradritti, Francesco and De Luca, Araldo. *Egyptian Treasures from the Egyptian Museum in Cairo.* Harry N. Abrams, Inc., 1999.

Waxman, Sharon. *Loot – The Battle Over the Stolen Treasures of the Ancient World.* Times Books, 2008.

Williams, A. R. *Modern Technology Reopens the Ancient Case of King Tut.* National Geographic Magazine, June 2005.

Abstract of the Article
Published in the *Journal Facial Plastic Surgery*
Dec. 4, 2018

Effects of Topical Mandelic Acid Treatment on Facial Skin Viscoelasticity

Stanley W. Jacobs, BSc, MSc, MD, FRCS(C)1 Eric J. Culbertson, MD2

1Department of Facial Plastic Surgery, The Jacobs Center for Cosmetic Surgery, Healdsburg, California

2Department of Plastic Surgery, The Jacobs Center for Cosmetic Surgery, Healdsburg, California

Address for correspondence Eric J. Culbertson, MD, Department of Plastic Surgery, The Jacobs Center for Cosmetic Surgery, 145 Foss Creek Circle, Healdsburg, CA 95448 (e-mail: eric@thejacobscenter.com).

Facial Plast Surg

Keywords
Cutometer
Skin elasticity
Mandelic acid
A-hydroxy acid
Antiaging

Mandelic acid is an α-hydroxy acid with reported benefit in treating acne and hyperpigmentation. The authors have developed a topical mandelic acid formulation that subjectively improves the quality of aged skin. Although the gold standard for assessing outcomes, photographic documentation is limited by subjective interpretation. Tools for measuring physical skin properties allow for an objective assessment of changes in skin quality. The authors sought to objectively study the viscoelastic changes to the skin followingtreatment with topicalmandelic acid, using theCutometerMPA580. Twenty-four patients, twenty females

Continued on next page

and four males, aged 42 to 68 years, were studied over a four week period. Mandelic acidwas applied topically to the face twice a day for four weeks. The lower eyelid skin viscoelastic properties were assessed weekly using the Cutometer. After four weeks of topical mandelic acid treatment, the elasticity of lower eyelid skin increased 25.4% (P < .003). Skin firmness increased 23.8% (P < .029). Improvement in photographic appearance correlated with these findings. Mandelic acid is another topical treatment option for improving skin quality, and is well tolerated by patients. The authors feel that the Cutometer or similar device should be used routinely in facial plastic surgery to objectively assess outcomes of various treatment modalities.

About the Authors

Stanley Jacobs, M.D., is a triple board certified facial plastic and reconstructive surgeon. He is certified by the American Board of Facial Plastic and Reconstructive Surgery, the American Board of Otolaryngology/Head and Neck Surgery and the by the Royal college of Physicians and Surgeons of Canada. He is Fellowship trained in facial plastic surgery at the University of California San Francisco. He did his general/cardiac surgery training and facial plastic /ENT training at the University of Toronto. He received his medical degree from the University of Western Ontario, in London Canada. He received his Master's degree in medical biophysics/immunology at the Ontario Cancer Institute, in Toronto. He received his Bachelor's degree in science with an honors in Anatomy, graduating with a Magna Cum Laude from McGill University in Montreal.

He has performed more than 12,000 facial procedures by the publishing of this book. He has been Chief of Surgery at Sutter hospital in Santa Rosa and Chief of Facial Plastic Surgery at Santa Rosa Memorial hospital, and has also served as the Specialty House Delegate for facial plastic surgery for the state of California. He has published and presented medical papers on skin elasticity, facelifts, chemical peels, and rhinoplasty worldwide.

Karen Hart has more than 30 years of experience as a writer and editor. She is the editor of *NorthBay biz* magazine, located in Northern California, and the author of the young adult novel, *Butterflies in May*. She has also been published in *The Press Democrat, Sonoma, North Bay Woman* and other publications. Her interest in ancient Egypt was ignited as a young girl when she first encountered an ankh, a heiro-glyphic symbol used in writing and art to symbolize life. Hart lives in Sonoma County.